This book, by way of in-depth analysis of multiple case studies across the world, offers fresh perspectives on the issue of conflict in coastal and marine areas. It reflects on the role of conflict, neatly brings together theory with examples of practice and presents a convincing rationale for why we need to engage with it to achieve management outcomes. Written in a very personal and engaging manner, the text immediately evokes the importance of the issues while retaining scholarly depth. This book will provide great insights into those interested in how to navigate the conflicts prevalent in our beautiful marine and coastal areas.

Melissa Nursey-Bray, University of Adelaide, Australia

Conflicts over Marine and Coastal Common Resources

This book explores the types of conflicts that occur over marine and coastal resources, the underlying causes, and attempts to prevent them.

Despite the emergence of various marine and coastal governance approaches to address the effects of human activities within the marine environment, conflict continues. In this book, the author outlines the reasons conflicts can, and do, arise in the marine and coastal environment. Drawing on case studies from both the northern and southern hemispheres, the book takes a broad view of how we interact with our environment, of how and why conflict is perpetuated as a political and cultural phenomenon, and how this varies or remains constant across space and place. The case studies explore not only the underlying perceptions and needs of those involved in marine and coastal conflict and the types of conflicts that arise in oceanic and coastal areas, but also the underpinning reasons for these conflicts. Marine and coastal resource conflicts have the potential to derail conservation efforts and blue growth policies, as well as the United Nations Sustainable Development Goals. Thus, it is imperative we understand the drivers and exacerbating factors of marine and coastal conflict. Arguing that there is an urgent need for renewed thinking and focus on conflict prevention, the author develops a theory of marine and coastal conflict which allows us to understand those factors and the means to help prevent such conflicts arising in the first place.

This book will be of interest to students and researchers of coastal and marine science and environmental management as well as those working in the field of marine resource management, including coastal zone managers and fisheries managers.

Karen A. Alexander is a research fellow in the Centre for Marine Socio-ecology, and the Institute of Marine and Antarctic Studies, at the University of Tasmania, Australia. She is a human geographer with wide-ranging interests, centring on marine governance. Karen specialises in issues around the transition to a blue economy and her research has focused on marine and coastal conflict, societal support for sectors such as offshore energy and aquaculture, coastal ecosystem-based management, and marine spatial planning.

Earthscan Oceans

Citizen Science for Coastal and Marine Conservation
Edited by John A. Cigliano and Heidi L. Ballard

Ocean Energy
Governance Challenges for Wave and Tidal Stream Technologies
Edited by Glen Wright, Sandy Kerr and Kate Johnson

Offshore Energy and Marine Spatial Planning
Edited by Katherine Yates and Corey Bradshaw

Coral Reefs: Tourism, Conservation and Management
Edited by Bruce Prideaux and Anja Pabel

Marine Extremes
Ocean Safety, Marine Health and the Blue Economy
Edited by Erika Techera and Gundula Winter

Marine Policy
An Introduction to Governance and International Law of the Oceans
2nd Edition
Mark Zacharias and Jeff Ardon

Conflicts over Marine and Coastal Common Resources
Causes, Governance, and Prevention
Karen A. Alexander

Marine and Fisheries Policies in Latin America
A Comparison of Selected Countries
Edited by Manuel Ruiz Muller, Rodrigo Oyandel and Bruno Monteferri Siles

For further details please visit the series page on the Routledge website: www.routledge.com/books/series/ECOCE

Conflicts over Marine and Coastal Common Resources

Causes, Governance, and Prevention

Karen A. Alexander

Routledge
Taylor & Francis Group
LONDON AND NEW YORK

earthscan
from Routledge

First published 2020
by Routledge
2 Park Square, Milton Park, Abingdon, Oxon OX14 4RN

and by Routledge
605 Third Avenue, New York, NY 10017

First issued in paperback 2021

Routledge is an imprint of the Taylor & Francis Group, an informa business

British Library Cataloguing-in-Publication Data
A catalogue record for this book is available from the British Library

Library of Congress Cataloging-in-Publication Data
A catalog record has been requested for this book

ISBN 13: 978-0-367-78537-6 (pbk)
ISBN 13: 978-1-138-63525-8 (hbk)

Typeset in Sabon
by Wearset Ltd, Boldon, Tyne and Wear

Contents

Illustrations

Figures

Tables

Foreword

Just over 50 years ago on 24 December 1968, Apollo VIII astronaut William Anders captured one of the twentieth century's most evocative images – the photograph entitled 'Earth Rise' – the earth rising above the surface of the Moon as the Apollo capsule continued its lunar orbit. Earth Rise's striking image emphasised the Earth's vulnerability against the backdrop of the infinite blackness of space. The photograph highlighted, too, for the first time Earth as the blue planet, with the oceans that cover 70 per cent of the planet's surface clearly visible and dominating the image.

At the same time as the Earth Rise image was taken the global community was urged to take focused action to the state of the planet's seas and oceans. After centuries of use, technological advances were raising concern over the way in which marine resources were being exploited. Significant incidents and accidents led to major marine pollution events; the grounding and wrecking of the oil tanker the SS *Torrey Canyon* off Cornwall in March 1967 and the Santa Barbara oil spill in early 1969. In November 1967, in response to these broadening concerns over perceptions of uncontrolled development of the ocean, Arvid Pardo, the Ambassador for Malta to the United Nations urged the United Nations General Assembly to address the management of the world's seas and oceans. Pardo's declaration that the seabed was 'the common heritage of mankind' provided the intellectual catalyst for a key concept contained within the United Nations Law of the Sea Convention. Allocating access to areas of commons and providing this access fairly and equitably was a key to Pardo's concerns. The development of the United Nations Law of the Sea Convention – a direct outcome of Pardo's speech to the General Assembly – provided what has been described as a constitution for the oceans. National governments, too, acted, reforming and extending regulatory frameworks for ocean and coastal management.

This has led to a complex governance framework, dominated by a sectoral focus to regulation and policy development but increasingly intersecting with market-based tools or instruments and community values and interests. Increasingly, too, the limits and challenges to sectoral approaches to marine and coastal management have become clear. Sectoral management can be effective but can intensify problems and conflicts. Identifying and managing conflict is a key to successful policy development. Decisions over coastal and marine issues are essentially political in the sense that they are about managing interests and allocating resources. A focus on conflict and its management – both in the positive sense of the articulation of different values and interests as well as the common negative connotation – goes to the core of politics as 'who gets what, when, how and why?'

Karen Alexander's *Conflicts over Marine and Coastal Common Resources* provides an important contribution to the study of marine and coastal conflict. Drawing on several illustrative case studies, this work revisits the challenge of managing marine and coastal areas as commons. It also draws on a range of insights into the study of conflict. This provides a strong background to the analysis of marine and coastal conflicts, and the development of a theoretical framework that recognises that a key factor is competition over goods and services (including non-use ecosystem services), the rights and interests of those affected, and the rules used to make decisions. *Conflicts over Marine and Coastal Common Resources* is a valuable contribution to the analysis and understanding of the diverse values and interests underpinning management of coastal and marine areas. Karen Alexander's work is timely and deserves to be widely read. As she notes, the increased emphasis on 'blue growth' is likely to increase utilisation of our marine and coastal areas. Understanding how to manage and accommodate different interests and values (that is centring on addressing and managing conflict) is as important as addressing the demands for, or capacities to, address blue growth.

Marcus Haward
University of Tasmania

Preface and acknowledgements

I grew up in a small community in a remote part of west-coast Scotland. Population 200, no street-lights. A beautiful location, the village of Kilchrenan sits in the shadow of Ben Cruachan and on the banks of Loch Awe – the longest freshwater loch in Scotland. Wildlife is everywhere, including iconic species such as red deer and golden eagles. It is a tranquil place, at its busiest on the infrequent Monday mornings that residents congregate in the village hall for coffee and cake. When I was in my teens, for a short time, the peace was shattered. A developer, GreenPower (Carraig Gheal) Ltd, submitted a proposal for a wind farm near our village. Almost immediately, our village was divided. Some saw the need for more renewable forms of energy and believed that the wind farm would benefit the community through the creation of a community fund. Others thought it would be an eyesore, would hamper tourism, and would kill migrating birds. Community meetings became shouting matches, friends stopped talking to friends. I could not understand what was going on. I didn't find the turbines that loathsome, I thought they were rather elegant. That was my first experience of natural resource conflict, and I was engrossed. I left to go to university to study. For a while, I forgot about my interest in natural resource conflict, focusing more on topics such as modern political thought and philosophy. But while studying a Master of Science in Environmental Governance, discussing conflicts over blood diamonds and deforestation, my interest was re-awakened. I have been researching various aspects of marine environmental resource conflict ever since.

Why should we care about conflict over marine and coastal resources? In the current era of the Anthropocene, when changes in the environment and the resources it provides are being greatly impacted on by human influences, the only way that we can ensure these resources survive into the future is by managing humans. The existence of conflict may suggest that we are not doing this optimally. If we can understand why conflict arises, and what we can do to prevent, reduce, or resolve it, perhaps we can identify how better to manage our marine and coastal environment.

Perhaps you make your living from, or recreationally use, marine resources. It may be that you manage ocean and coastal controversy and conflict. Possibly, you are interested in marine and coastal matters, or in natural resource conflict. Whichever it is, I believe that if you care about the blue planet on which we rely (or even one small part of it), recognising and understanding the tensions and challenges arising over how to use, conserve, and manage our oceans and coasts will be crucial for you.

I hope that the approach taken in this book provides information that will help marine and coastal decision-makers, and others, to grasp the issues, to start to think differently about marine conflict-related problems, and to make effective decisions to prevent conflict in the future. The contentious issues confronting marine management today are sufficiently urgent to demand better explanation and improved governance options and to do this we need to dig deeply into the subject. For those willing to invest the time in understanding these issues, your investment will be rewarded.

No woman is an island, and several people were instrumental in the writing of this book. Thanks, and blame, goes to Meredith Nash for putting the idea of writing a book into my head in the first place. Thanks particularly to Abby Cvitanovic (and family) for the early-doors support – Abby, I still have your note on my fridge! Thanks also to many of my colleagues in the Centre for Marine Socioecology in Tasmania for influential conversations and ongoing support, particularly Emily Ogier (Em, I bet you are glad you don't have to hear about the 'buke' anymore!), Peat Leith, Maree Fudge, Kirsty Nash, and Gretta Pecl. Hugely important reviews of early drafts were undertaken by Marcus Haward and Jo Vince and editing by Colin Evans and John Bainbridge – you all have my eternal heartfelt gratitude. Thanks also to my fabulous editors, Tim Hardwick and Amy Johnston. As ever, my parents were extremely supportive right up until the last minute – I couldn't have done this without you.

Should you be interested, the Carraig Gheal wind farm was approved. The community now benefits from revenue from the wind farm with investment in the village hall and acquisitions such as a defibrillator – a necessary addition to a village with an average age of 60.

Acronyms and abbreviations

AFMA	Australian Fisheries Management Authority
ANA	Asamblea Nacional Ambientalista
BC	British Columbia
BIM	Bord Iascaigh Mhara
BP	British Petroleum
CBCRM	Community-based coastal resource management
CBD	Convention on Biological Diversity
CBSFA	Community-based subsistence fishing area
C-CONDEM	Coordinating Committee for the Defence of Mangroves
CMT	Customary marine tenure
CO_2	Carbon dioxide
DWT	Devon Wildlife Trust
EEZ	Exclusive Economic Zone
ENGO	Environmental non-governmental organisation
EPA	Environment Protection Agency
EU	European Union
FAO	Food and Agriculture Organisation of the United Nations
FUNDECOL	Foundation of Ecological Defence
GVA	Gross value added
HELCOM	Helsinki Convention for the Baltic Sea
IDPPE	National Institute for the Development of Small-scale Fisheries
IPCC	Intergovernmental Panel on Climate Change
ITQ	Individual transferable quota
LMMA	Locally-managed marine areas
MARPOL	International Convention for the Prevention of Pollution from Ships
MEP	Member of the European Parliament
MGR	Marine genetic resource
MPA	Marine protected area
MSP	Marine spatial planning
MW	Mega-watts

NGO	Non-governmental organisation
NIMBY	Not in my backyard
NM	Nautical mile
NMP	National Marine Plan
NZ	New Zealand
OECD	Organisation for Economic Co-operation and Development
OSPAR	The Convention for the Protection of the Marine Environment of the North-East Atlantic
PRC	People's Republic of China
PRIO	International Peace Research Institute in Oslo
SAMP	Special area management plan
SANParks	South Africa National Parks Board
SDGs	Sustainable Development Goals
SSMEI	Scottish Sustainable Marine Environment Initiative
TAC	Total allowable catch
TAE	Total allowable effort
TURFs	Territorial use rights in fishing
UNCLOS	United Nations Convention on the Law of the Sea
UNESCO	The United Nations Educational, Scientific and Cultural Organization
UNFCCC	United Nations Framework Convention on Climate Change

Chapter 1

All at sea

Conflicts regarding governance of the marine environment, particularly where multiple stakeholders are involved, are many and varied. Such conflict is typically based on a complex cocktail of legacy and assumed future outcomes as perceived by the many users of, and dependents on, the marine ecosystem in question. The results of extensive debates, research, and assessments do not necessarily result in the balanced use of marine resources. Moreover, often, insufficient mechanisms exist to ensure that all social, economic, and environmental elements are considered in moving to a dynamic equilibrium. In this book, I am going to take you on a journey where we explore why marine and coastal conflict occurs and if, and how, our existing marine governance mechanisms can contribute to preventing such conflict. But first, let's get a feel for what sort of conflict I am referring to.

As the United Kingdom raced to meet a European Union (EU) goal of 20 per cent of the EU's energy needs from renewables by 2020, the Crown Estate (managers of the UK seabed) ran several offshore wind leasing rounds. In 2009, as part of the second round, Centrica submitted a planning consent application for a wind farm at Race Bank, off the coast of Norfolk, covering an area of $75\,km^2$. By November 2013 the project's future was in doubt as it failed to receive subsidies from the UK Department of Energy and Climate Change. In December 2013, the project was sold to Dong Energy. In 2014, when fishers were forced out of their traditional fishing area, a conflict arose between the fishers and the company. Dong Energy (majority owned by the Danish Government) had secured an injunction that gave fishers less than five days to remove thousands of crab-fishing pots from the lease area in the Greater Wash. Fishers believed that the compensation Dong had offered them was not fair and argued that Dong Energy's 'negotiating style stinks' (Gosden, 2014). They felt that Dong had given insufficient time to reach an agreement and, when they had not got what they desired, had resorted to bullying tactics. Nicky King, 51, chairman of the Wells and District Inshore Fisherman's Association, said: 'We understand we're not going to prevent a multi-billion-pound project going ahead but we feel we should be able to receive a fair compensation' (Lazzari, 2014). Fishers perceived that a large foreign company had tried to force them out of the area they had fished in for generations. They felt that 'at the moment money talks and might is right' (ibid.). The conflict was resolved in 2015 as the fishers and Dong Energy came to an agreement over compensation, although neither side revealed the size of the new offer.

A plan initiated in 2011 to create a marine reserve off Waiheke Island in New Zealand divided an island community. A report released by the Hauraki Gulf Forum, suggesting that the Hauraki Gulf was suffering from serious environmental degradation and a loss of biodiversity, led to the creation of a group called 'Friends of the Hauraki Gulf' whose aim was to drive the proposal for a marine reserve. However, push-back came from nature-loving locals who were worried about losing access to *kaimoana* (seafood) as well as the potential for increased crowding and commercial activity due to elevated tourist numbers visiting the reserve. 'The issue, basically, is that this is the wrong place for a reserve. Everybody would love to have reserves, but not there. This is our larder. This is our food source', Danny Shortland from Piritahi Marae told the *Sunday Star Times* (Johnson, 2013). Mostly, the two opposing camps staked out their positions in letters to the editor and paid newspaper ads, at local council meetings and in media interviews. Tensions were also highlighted when Parliament considered a petition raised by a group called 'Keep our Beaches', although Ministers of Parliament concluded that as no application for a marine reserve had been lodged it was inappropriate to intervene.

Two years later, in 2013, the Norwegian government decided to halt oil and gas drilling around the Lofoten and Vesteraalen islands in the Norwegian Arctic. The seas off the islands are home to the spawning grounds of the world's largest cod stock, leading fishermen and environmentalists to fear that oil activities could endanger the stocks, thus pitting them against the oil industry. Much of the battle was played out in government with attempts to give priority to the oil industry being blocked by opposing parties in the Storting (parliament) and in 2013 the newly elected Conservative Party formed a minority government with the backing of other parties based on the promise that they would protect the Lofoten Islands. Nevertheless, the substantial resources in the area – an estimated 1.3 billion barrels of oil in Lofoten and neighbouring Vesteralen and Senja – continues to fuel new political debate. As of 2018, the oil-friendly government coalition of the Conservative Party and the Progress Party have been unable to proceed with plans for impact studies in the area because of opposition from the Liberals and Christian Democrats.

In 2013, around 2000 people marched through Galway City in the Republic of Ireland bearing banners with the names of more than 20 angling and protest groups. The march was in response to an application by Bord Iascaigh Mhara (BIM) for a licence to farm 15,000 tonnes of salmon in an area covering 1,126 acres which, it suggested, would benefit the local community through a €60 million initial investment plus an annual wage flow of over €14 million. The proposal did not have the support from the local communities and stakeholders who believed it would affect angling, fisheries, and tourism in the area. Local Aran islander Ruari O Cualain told *BBC News*:

> We've a beautiful beach, we've beautiful clean waters. Who is going to visit the place with a huge fish farm just a mile off it, with noise there 24 hours a day, with smell from fish? I don't even know if I want to live here if the fish farm comes in here.
>
> (Fleming, 2013)

Following the protests, the marchers sent a delegation to the International Skipper Expo where they handed in a letter to personnel on the BIM stand. Non-governmental organisations (NGOs) lodged complaints with the official government watchdog and suggested that if the project was approved they could bring a legal challenge. Even government agencies were at loggerheads over the project: while the fisheries board proposed the project, Inland Fisheries Ireland – the country's authority for recreational fishing – was against it. In 2015, BIM withdrew the application to develop the world's largest offshore salmon farm off the west coast of Ireland.

These stories are four examples from among hundreds. They share two key components. First, each describes proposed changes to the way that humans currently use the ocean and coastal environment, be it for development or conservation purposes. Second, each describes a situation which became engulfed in controversy and conflict.

Unmanaged conflict can affect us in a variety of ways and at a variety of scales. It has the potential to cause negative consequences for individuals. This may include increased stress, lowered productivity and motivation, damaged emotional and psychological wellbeing, mistrust, and hostility (which can turn, and has on occasion turned, to violence). At the organisation level, conflict may lead to wasted time and money, as well as poor decision-making. For society, a result of unmanaged conflict could be economic downturn or even the breakdown of social structure.

Conflict can also have a positive effect. Conflict can encourage new thinking; the consideration of different points of view. Conflict can lead to new questions being asked, it opens minds and beats stagnation. Conflict can also help us to find common ground, foster intellectual, emotional, and moral growth, and can foster change. But are we managing conflict in the marine and coastal environment? Are we harvesting its potential for positive outcomes? The evidence would suggest not.

Time for an intervention

In the early decades of the twenty-first century, there was rapid population growth and substantial technological advancement. In early history, fishers used barbed poles to spearfish or used woven baskets and nets, often in small boats close to the shore. Today, large factory trawler vessels can cross oceans. With a growing demand for high quality protein, and a decrease in many of the world's fish stocks, we have seen the development of modern aquaculture, which has included mariculture. As fossil fuels become increasingly harder to access, the need for energy security has led to technological developments that allow the harvest of renewable energy from tidal streams and wave energy. While recreational use of the sea has occurred for thousands of years, as prosperity has increased so has the number of recreational cruising and racing vessels in recent years. This has occurred alongside a growing interest in marine ecotourism, which includes such activities as marine mammal watching and adventure expeditions to the poles. The importance of the oceans and coasts for economic growth became clear in the concept of the 'blue economy' (linked to the concept of 'blue growth'), which came out of the 2012 Rio20+ Conference, and which has become a key development policy for many of the world's coastal nations.

At the same time, there has been an increasing push to protect and preserve our marine environment. Human actions unavoidably modify or transform the oceans and coasts. In some cases, this has led to problems such as coastal erosion, loss of habitat, extinct and endangered species, and increases in foreign and invasive species. As a result, new conservation tools have been created such as marine reserves and marine protected areas (MPAs), which are areas of the sea where human activity is restricted for conservation. We may choose to limit development, fishing practices, or moorings in these locations. MPAs can also provide revenue from more passive human activities such as marine tourism. Many nations around the world have begun to allocate areas of sea space to MPAs; indeed, Target 11 of the Aichi Biodiversity Targets aims to protect at least 10 per cent of coastal and marine areas globally.

The drive for development and conservation is framed by the Sustainable Development Goals (SDGs), an interlinked set of 17 goals set by the United Nations General Assembly as part of Agenda 2030. The SDGs aim to address many of the global challenges we currently face, including those related to poverty, inequality, climate, environmental degradation, prosperity, and peace and justice. Conflict prevention or resolution, particularly between development and conservation, is a pre-requisite for achieving these goals. In fact, increasing maritime conflict and related security issues are a 'quadruple bind' because they put income security (SDG 1), food security (SDG 2), energy security (SDG 7), and safety

(SDG 16) at risk. Moreover, these issues are also likely to affect SDG 14 (life below water) and SDG 15 (life on land). With the global drive to achieve sustainable development, we are likely to see increasing conflict. Perhaps now is the time for an intervention.

Yet, there is a gap in our understanding of why conflict arises in the marine space in the first place. An oft-used, and occasionally throwaway, explanation for conflict (most often used by proponents of the development or conservation measure) is that of 'NIMBY (not in my backyard)-ism'. The concept of NIMBY suggests that an individual or organisation believes some sort of change is needed in society, but they do not want that change nearby them. The term is used as a pejorative and, although in some cases this may be one reason for the raised interest of a group it is seldom the sole catalyst for opposition to any proposed change in how we use our oceans and coasts. It is unacceptable to work on the sole assumption that NIMBY-ism is the reason for opposition to any proposed change in how we use our oceans and coasts.

Those who think more critically about such issues often propose that the reason for these conflicts is that marine users are fighting over space. That if we just understand which areas are used and valued, then we can figure out how to resolve these conflicts. Indeed, this has led to an increased focus on approaches such as marine spatial planning (MSP; also known as ocean zoning although it is much more than that), the uptake of which has been substantial. Again, it is hard to believe that this is the sole explanation for why conflict occurs in this space.

In fact, there are many diverse interests at play in the ocean and coastal environment, and as such there are likely to be a range of causes of contestation. Therefore, to address contentious issues, such as those described in the earlier stories, decision-making must be adaptive and, more importantly, informed. We must fully understand why these conflicts arise and use this information to promote change in ocean and coastal governance to ensure that such conflicts do not subvert our ability to achieve the SDGs.

Aim and structure of the book

In '*Conflicts over marine and coastal "common" resources*', I aim to discern conflict and contestation in the marine and coastal environment, what it means for sustainable development (as understood through the SDGs), conflict theory and practice. To do this, I take a broader view of how we interact with our environment, of how and why conflict is perpetuated as a political and cultural phenomenon, and how this varies or remains constant across space and place. Like many human geographers, in this study I draw on research from such diverse fields as environmental science, economics, sociology, political science, and psychology.

As we will see, no one existing theory provides an overarching explanation for why conflict occurs in the ocean and coastal environment. Consequently, governance mechanisms which attempt to prevent marine conflict are built on an inadequate theoretical foundation. This means that they may do more harm than good. Rather, the problem is that there is no single explanation, but several partial explanations. In this book I tie together these explanations with new case study analysis to develop a theory of marine and coastal conflict. I argue that increased competition for ocean and coastal goods and services, driven by a changing social-ecological environment, may be transformed into conflict by political and social factors that shape the rules, rights, and effects of human resource use. Furthermore, I believe that by understanding the drivers and exacerbating factors of marine and coastal conflict, we can adapt marine governance processes in ways which may help to prevent such conflicts arising in the first place.

A substantial component of this book is the analysis of marine and coastal conflict case studies from around the globe. To do this, I have used discourse analysis (Gee, 2004) to analyse and interpret written communications (such as academic journal articles, court reports, industry and NGO literature, traditional and social media articles) relating to several marine and coastal conflicts. My methods comprised data reduction, data display, and conclusion drawing (Miles *et al.*, 1994) using QSR International's NVivo10 software (QSR International Pty Ltd, 2012), and a grounded theory approach to analysis (Charmaz, 2006). I then use the theory developed during this analysis as a frame to critically appraise how some existing marine governance structures address (or not) the underlying causes of marine and coastal conflict.

This chapter has introduced the topic of interest and provided some idea of what the book will cover. Chapter 2 identifies some peculiar characteristics of the ocean and coastal environment, providing some background as to why we may need to consider terrestrial and marine conflict differently. Chapter 3 delves into some of the existing theories of conflict and critically assesses how useful these are in the context of the oceans and coasts. In Chapter 4 I introduce the case studies that I analysed to develop a new understanding of marine and coastal conflict. Chapters 5 and 6 introduce the key drivers of ocean and coastal conflict, and the factors that may cause these conflicts escalate. Based on these findings, Chapter 7 presents a new theory of ocean and coastal conflict and compares this to existing theories. In Chapters 8 to 10 I review three existing means of preventing ocean and coastal conflict – legislation, MSP, and co-management – analysing if and how they address the exacerbating factors identified in Chapter 6. Finally, Chapter 11 provides a summary of the findings and the implications of these findings.

References

Charmaz, K. 2006. *Constructing grounded theory: A practical guide through qualitative analysis*. Sage.

Fleming, D. 2013. *Irish battle over plans for Aran giant salmon farm* [Online]. BBC News. Available: www.bbc.co.uk/news/world-europe-23550159 [Accessed 27 February 2017].

Gee, J. P. 2004. *An introduction to discourse analysis: Theory and method*. Routledge.

Gosden, E. 2014. Norfolk crab fishermen forced out to make way for offshore wind farm works. *Telegraph*, 13 August.

Johnson, S. 2013. The battle of Waiheke Island. *Sunday Star Times*, 1 September.

Lazzari, A. 2014. Fishermens' anger as offshore wind company issues injunction to prevent fishing in an area of The Wash. *Eastern Daily Press*, 15 August.

Miles, M. B., Huberman, A. M., Huberman, M. A., and Huberman, M. 1994. *Qualitative data analysis: An expanded sourcebook*. Sage.

QSR International Pty Ltd 2012. NVivo qualitative data analysis software version 10.

Chapter 2

The marine and coastal 'commons'

Before we begin to consider the notion of conflict, it is worth our taking some time to reflect on the environment in which this conflict takes place – the oceans and coasts. Such context is critical to our understanding because the marine environment is substantially different to the terrestrial. It is a different physical space, it holds very different natural resources, it is governed differently, and what is more, we perceive it in a very different way. It is for this reason that a separate treatment of marine and coastal conflict is required in the first place.

Let's start with some definitions. The ocean, also known as the marine environment, is the vast body of saltwater covering 71 per cent of the earth's surface. It provides habitat for a vast resource of living organisms, from the smallest bacteria to the largest whale; is the main source of oxygen (from phytoplankton); and a major reservoir of carbon dioxide (CO_2). The coast, however, is slightly more difficult to describe and define. When I talk about the coasts, I am referring to the World Bank definition:

> ... the interface where the land meets the ocean, encompassing shoreline environments as well as adjacent coastal waters. Its components can include river deltas, coastal plains, wetlands, beaches and dunes, reefs, mangrove forests, lagoons, other coastal features.
>
> (Post and Lundin, 1996)

In this book, I examine conflict in both spaces, but specifically in the coastal zone, out to 200 nautical miles (nm) of a nation's Exclusive Economic Zone (the area in which a country claims exclusive rights for fishing, drilling, and other economic activities; Figure 2.1). Combined, these areas comprise well over three-quarters of our world – making them central to all our lives.

In this chapter, I introduce the key characteristics of the ocean and coastal environment. In some depth, I explore the ways in which it is different from the terrestrial, for this is a key concept underpinning the need for a separate treatment of ocean and coastal conflict. I ask what

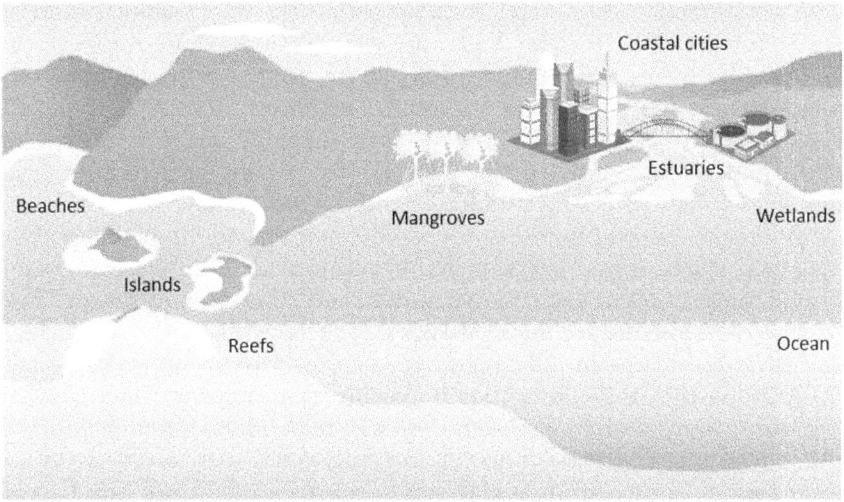

Figure 2.1 The marine and coastal area; the coastal zone.

Source: Courtesy of the Integration and Application Network, University of Maryland Centre for Environmental Science (ian.umces.edu/symbols/).

makes the ocean and coastal environment unique. To answer, I examine the material environment, the ecosystem services (the benefits we obtain from the environment) it provides and their 'common goods' nature, the different governing conditions and the 'contested spaces' aspect of this specific environment.

The abiotic environment

Several environmental features are peculiar to the oceans and coasts, and influence the use of these areas. First, we tend to consider use of terrestrial space in two dimensions, in terms of width and length. Although we have airspace above land and subterranean area below, most land use (such as farming, forestry, or hiking) occurs on the surface, in that two-dimensional space. On the other hand, water space occupies three dimensions: width, length, and depth. We use the surface of the water for e.g. boating (which covers the width and length aspects) and we use the water column (the depth aspect) for fishing. In addition, we have the 'secondary' surface of the seabed which we use for mining, dredging, etc. This adds an extra layer of complexity, particularly as we cannot immediately see the depth or the secondary surface components.

Terrestrial environments also have more pronounced physical boundaries than do marine areas (Thorne-Miller, 1999). We can clearly see the boundaries between tundra, grasslands, tropical rainforests, and deserts, and land-based species tend to be tightly tied to specific terrestrial environments. Due to the fluid and dynamic nature of marine ecosystems, however, the boundaries are more 'fuzzy' or transitional and many species are not bound to specific locations. This is particularly clear in terms of the central role of larval dispersal and connectivity in sustaining marine populations. The fluidity of this environment also means that nutrients and pollutants can spread widely, sometimes on short timescales, cutting down response times to environmental and/or human perturbations. For example, it takes only ten minutes for a spill of one tonne of oil to disperse over a radius of 50 metres, forming a slick of up to ten millimetres thick. Waves and turbulence on the sea surface can then break up the oil into fragments and droplets and it spreads throughout this three-dimensional marine environment (International Tanker Owners Pollution Federation Limited). Human activity in the ocean and coastal area is also less spatially fixed than on land. As an example, while we can only play some sports in a field, we can dive or snorkel almost anywhere in the sea.

Finally, temporal variations on daily, monthly, and seasonal bases are also more prominent in the oceans than on land, and human activity in the marine environment often reflects these cycles. Fisheries may close at spawning times to protect breeding stock or to limit the catch for a fishery. Public health agencies may also close a fishery at certain times of the year to prevent the ingestion of biotoxins from shellfish. And, in some locations, recreational boating and surfing are more common in summer when the weather, and water temperatures, are more amenable.

Marine and coastal goods and services

Another way in which the terrestrial and marine environments are different is that they provide very different ecosystem goods and services, the benefits of which accrue to all living organisms including plants, animals, and humans. These goods and services are extremely important for human society in terms of health, economic, and cultural needs, and it is most often over these goods and services that conflict arises.

According to the Millennium Ecosystem Assessment, people obtain four key services from ecosystems: supporting services, regulating services, provisioning services, and cultural services (Sarukhán *et al.*, 2005). While these services are found in terrestrial and marine environments, there are some key differences. For example, certain plant and animal life is uniquely designed to thrive under the physical and chemical conditions of the oceans and coasts. Let us consider how the marine environment provides each of these four types of ecosystem services.

Supporting services

Supporting services maintain the conditions for life on Earth. Their impacts on people are either indirect or occur over long timescales.

A key supporting service provided by the ocean is primary productivity (the synthesis of new organic material from inorganic materials such as water and CO_2 in the form of algae). There have been many different estimates of the total amount of primary production in the ocean, but the consensus is that the value is about 45–50 Gigatons of Carbon per year, just under half of total annual net primary production (Longhurst *et al.*, 1995). Almost all life on Earth relies directly or indirectly on primary production, with microscopic algae providing food for microscopic animal plankton, which is food for small fish and animals, which larger fish, birds, and humans then eat.

Marine and coastal areas are also part of a recycling system which delivers vital ecosystem components such as carbon, nitrogen, and phosphorous, back into the environment for further use. These areas are also vital to the Earth's water cycle, which connects the oceans, the land, and the atmosphere, and is responsible for distributing water across the Earth through evaporation, condensation, and precipitation, thus sustaining life by providing water to plants, animals, and humans. It is also the driving force behind the Earth's climate. Condensation forms clouds, and evaporation and water vapour in the air regulate temperature.

Regulating services

Regulating services are the benefits that humans obtain from the regulation of ecosystem processes. They are often invisible and so tend to be taken for granted. However, if damaged, the resulting losses could be substantial and difficult to restore.

Marine and coastal environments provide carbon sequestration and climate regulation services. Carbon sequestration is the process involved in carbon capture and the long-term storage of atmospheric carbon dioxide. The oceans are the main store of carbon dioxide (CO_2), and are believed to have taken up approximately 40 per cent of anthropogenic-sourced CO_2 from the atmosphere since the beginning of the industrial revolution (Reid *et al.*, 2009). The oceans also play a role in climate regulation especially in buffering (neutralising) the effects of increasing levels of greenhouse gases in the atmosphere and rising global temperatures.

Waste breakdown and detoxification is another regulating service that the oceans and coasts provide. Wetlands are the main natural wastewater treatment mechanism, filtering both human and animal waste and acting as a natural buffer to the surrounding environment. Micro-organisms in the soil help to break down waste. The process also eliminates pathogens (disease causing microbes) and reduces levels of nutrients and pollution. In addition to wetlands, fish, filter-feeding molluscs, and other aquatic animals are essential for water quality.

Finally, marine and coastal areas moderate extreme events such as erosion, wave damage, tsunamis, storms, and flooding, and over the long-term, rising sea levels. Mangroves are the front line of defence due to their position relative to many coastal hazards. They modify coastlines through their ability to attenuate waves, capture sediments, and build soils, enabling them to reduce the strength of the impact on directly exposed land. Seagrass beds are stabilisers, trapping sand and sediments by slowing water movement, and can prevent erosion of beaches and shores from the action of waves, and help to keep water clear. Reefs (coral and rocky) also create a natural buffer or barrier against wind and waves that would otherwise erode coastlines and affect coastal communities and infrastructure, such as ports and boat moorings, canals and marina developments.

Provisioning services

Provisioning services are the products that we obtain from ecosystems. These are the services that most likely first spring to mind when we think of marine resources. Provisioning services are often traded in markets and, in many regions, individuals may directly depend upon these services for their livelihoods.

The most obvious of these provisioning services is the harvest and culture of fish and shellfish. A significant amount of aquatic animal protein contributes to the diets of people around the world. However, as of the twenty-first century, the level of use of capture fisheries (marine and freshwater) is not sustainable and many fisheries have already collapsed. One-quarter of important commercial fish stocks are overexploited or significantly depleted (Sanchirico and Wilen, 2007). This has led to the rise of the large-scale commercial aquaculture industry. Algae is also harvested and increasingly cultured. Coastal peoples have used different types of marine algae for food for centuries. Carrageenan is a red alga principally used as a thickening and stabilising agent. Agar, a jelly-like substance obtained from various species of algae, is used as an anti-drying agent in breads and pastries (to provide a longer shelf life), among other uses. The seafood sector derives crucial cash earnings and employment for many millions of people globally, both in primary industry and in secondary activities such as processing.

Around 2,000 species and millions of specimens are traded annually in the ornamental fish trade, divided almost equally between fresh and saltwater (Livengood and Chapman, 2007). Although exact figures do not exist, it is estimated that the value of ornamental fish and invertebrates imported into different countries worldwide is approximately US$278 million (ibid.). The aquarium trade relies predominantly on wild collected specimens, with over 90 per cent of the traded marine species harvested from coral reefs (ibid.).

Ocean and coastal ecosystems also provide a variety of plants and animals that are used in traditional medicine, such as starfish, sea urchins, and sea cucumbers. People in Asia have been using seahorses for thousands of years as a cure for a variety of ailments such as impotence, incontinence, and swelling. In India, piles, coughs, and tuberculosis have been treated using corals. Algae are also a rich source of beta-carotene and other carotenoids used widely in pharmaceuticals. There is an increasing interest in the use of marine genetic resources (MGRs) from the sea and MGRs have been used in pharmaceuticals (e.g. painkiller Prialt from cone snail *Conus Magus* and cancer drug Yondelis from the sea squirt *Ecteinascidia turbinata*). MGRs have also been used in cosmetics (e.g. Estee Lauder cosmetic cream from soft coral *Pseudopterogorgia elisabethae*). Worldwide, many patents have been submitted with the intent of using MGRs for a range of purposes.

Cultural services

Finally, ecosystems strongly influence human culture. The non-material benefits that we obtain from ecosystems are called 'cultural services', and these are tightly bound to human values and behaviour, as well as to human institutions and patterns of social, economic, and political organisation.

People often choose where to spend their time based, at least in part, on the characteristics of the natural environment. The oceans and coasts provide myriad opportunities for recreation including angling, diving, kayaking, marine tours, sailing, snorkelling, surfing, swimming, and walking. Many of these activities also provide the basis for marine tourism, which attracts millions of travellers worldwide. Coral reefs such as the Red Sea reefs in Egypt, the Florida Keys Reef in the US, and the Great Barrier Reef in Australia are particularly important for tourism.

Seascapes are often places we want to visit and view, photograph or paint. Oceans have inspired artists for centuries – from the rock art and paintings of Easter Island, Baja California, and Australia, to the marine creatures adorning the vases of the ancient Greeks and the frescoed tombs of Egyptian pharaohs. Underwater photography is increasingly common with the first commercially available underwater cameras released in the mid-twentieth century and the development of underwater digital technology in the early days of the twenty-first century.

Society has always been linked to aquatic systems and many religious and spiritual traditions emanate from this relationship. Deities relating to the marine environment can be found throughout the world, including Mazu the Chinese sea goddess who protected seafarers, Tangaroa, the

Māori god of the sea and the father of many sea creatures, the Hebrew sea monster Leviathan and perhaps the most well-known Kings of the sea Neptune (Roman) and Poseidon (Greek). Many important traditions and festivals are linked to annual fish harvesting cycles. Among them are aquatic culinary events such as the Whale Shark Festival held annually in the municipality of Donsol in the Philippines and the turtle calling ritual on Fiji's Koro Island.

Oceans and coasts also inspire creative thinking, reduce anxiety, and promote compassion. It is believed that when we are near, on, in, or under the sea, we get a cognitive break because less information is entering our brains than in our normal day-to-day state. It allows us to be in a more mindful and contemplative state. Indeed, individuals living near the coast have been found to be generally healthier and happier than those living inland, and this effect is strongest in more deprived coastal communities (Nichols, 2014).

Multi-faceted management

We are already getting a feel for just how complex the marine and coastal environment is, as well as our relationship with it. This complexity requires substantially different modes of governance and management.

Marine and coastal resources are considered 'common pool resources', resources that can be human (products of civilisation such as the internet) or natural (earth-system components such as the atmosphere). They are resources that are finite or renewable, but depletable. They are resources large enough to make it difficult, but not impossible, to define recognised users and to exclude other users altogether. Furthermore, they are resources in which each person's use can potentially remove from the benefits that others might enjoy.

The Tragedy of the Commons is often a starting point for discussions relating to the idea of common pool resources. Until the 1980s it was the principal way in which people thought about the commons. To explain this concept, Hardin (1968) used the example of an imaginary field to which cattle herders have free and open access (i.e. the field is a 'commons' in which the cattle can graze). Each herder receives direct benefit from adding one more cow to the field, whereas all who are using it share the cost of grazing. Therefore, each herder has the incentive to put as many cows in the field as he can, an economically rational choice, but it leads to the degradation of the field – hence 'the tragedy'. Many academics believe fisheries to be a classic example. The fluid and migratory nature of fish make it difficult to protect rights to fish in the sea and the rule of capture prevails, leading to overexploitation.

Between 1980 and 1990, however, researchers such as Elinor Ostrom and Fikret Berkes (2005) began to find many exceptions to the tragedy concept. These exceptions came from a variety of resource types including fisheries, forestry, irrigation, and ground water. This meant that a new theory of the commons was required. Researchers started by identifying that common property (common pool) resources shared two characteristics: (a) the exclusion of, or the control of, access to potential resources was difficult, and (b) in using a resource, each user could subtract from the welfare of all other users (Ostrom *et al.*, 1999). Second, researchers suggested that common property was held in one of four basic property regimes (Berkes, 2005). In an open-access regime, access is free and open to all. In private property an individual or organisation has the right to exclude others and to regulate the use of a resource. State property access and usage is controlled by government. In communal property regimes, similar control and regulation is held by a community of users. Berkes (2005) goes further to note that these four regimes are ideal types but in practice, resources are usually held in a combination of the four and this is certainly the case in the oceans and coasts.

There are also several other governing considerations. As the terrestrial environment is effectively two-dimensional, this leads to single land use. This might be in relation to residential, commercial, or industrial areas, but we tend to see only one use per parcel of land. Aeroplanes may fly overhead, but they will not interact with land use. Similarly, pipes or cables may run under the surface of the land but will not interact with land use. Generally, we do not see any interaction between the different land uses. In the marine environment, however, we see that use of sea area is not exclusive because of its three-dimensional characteristic. It is a multi-use environment. As an example, telecommunications cables may be laid on the seafloor, fishers may fish in the water column and recreational vessels may sail on the surface – all in the same geographical location. These different uses and activities may also interact with each other. Cables may aggregate certain fish and shellfish species. Fishers may catch these species, but they may also snag their fishing gear on the cables or on rock armouring used as cable protection. Vessels sailing overhead may also prevent divers from surfacing. The marine environment is a complex and interactive one with substantially more potential for conflict.

On land, there are also private property rights with absolute land ownership, whereas in the sea there are quasi-property rights. We have a well-established market in terrestrial property rights where the associated obligations are codified legally or culturally. Indeed, the regional human rights instruments of Europe, Africa, and the Americas recognise the right to, and protection of property to varying degrees. Rights in the marine environment, on the other hand, are currently in a process of transition. Increasing resource use is creating a demand for private property rights as users lay claim to resources, and conflict is arising over who can lay claim to these resources. This started with the creation of sovereign rights. The

United Nations Convention on the Law of the Sea (UNCLOS), which came into force in 1994, defines the rights and responsibilities of nations with respect to their use of the world's oceans, establishing guidelines for businesses, the environment, and the management of marine natural resources. It set up rights with a 12 nm limit of the mean low water mark (territorial sea) and rights within the 200 nm limit (the exclusive economic zone). New types of private rights have also been, and are being, established such as fishing quota, which set a species-specific total allowable catch (TAC), typically by weight and for a given time. Seabed leases for oil and gas extraction, aquaculture, and marine renewable energy extraction have also been established.

Third, urban development drives governance of the terrestrial environment whereas conservation tends to drive marine governance. Land-based decision-making often occurs on local scales, with local authority planning departments responsible for development. Although predominantly concerned with the planning of settlements and communities, development of water use and resources, rural and agricultural land, parks and conserving areas of natural environmental significance are also governed at this level. Marine-based decision-making is generally much more centralised, with State or Federal government agencies at the helm. As an example, in Scotland, UK, responsibility for the marine environment to 200 nm mostly lies with Marine Scotland, a civil service directorate within the Scottish Government. In Australia, control of the marine environment out to 3 nm lies with the relevant State government (across a variety of departments) and from 3 to 200 nm lies with Federal government. This centralised focus takes decision-making power away from the localised area.

Last, due to its physical conditions and the spatial focus of humans, we have a more comprehensive understanding of the terrestrial environment. For the same reasons, we have an incomplete knowledge of the marine environment. Although we have not explored every corner of the planet's surface, the unexplored areas mostly comprise desert, mountains, deep caves, and the poles. The terrestrial environment is mostly easy to access and therefore to study and to monitor. In contrast, by the early twenty-first century, we have explored less than 5 per cent of the world's oceans (National Oceanic and Atmospheric Administration). Exploration and understanding of the marine environment is much more time-consuming, difficult, and costly. Its size alone is prohibitive. Add to that problems of weather, of pressure, and temperature, and it is understandable why we know so little about this vast domain. This has implications for marine and coastal conflict because knowing so little about this environment means that we do not fully understand how people affect it, nor how changes to it affect people and the long-term implications of decisions.

Contested spaces

It is likely that we can all easily conjure up a mental image of a terrestrial area which holds some deep personal meaning – a mountain ridge, an urban park, or a city centre shopping mall. Similarly, many of us can probably recall a stretch of coastline or a seascape. People develop a complex fabric of meanings of, relationships to and interactions with different locations. However, the ways in which we conceptualise these locations differ between terrestrial and marine environments.

In the mercantile era (between the sixteenth and eighteenth centuries), sea was central to life on land. There was substantial colonial expansion, particularly by northern European nations. Many nations expended significant effort to conquer new lands that would become sources of raw materials such as gold or sugar, as well as exclusive markets with monopolies operating in certain areas such as the Dutch East India Company and the Hudson's Bay Company. It was at this time that Hugo Grotius, who postulated that the sea was common property to all, and all nations should be free to use it for seafaring trade, published the book *Mare liberum* (Freedom of the Seas) (Grotius, 1916).

This changed in the late eighteenth century when the terrestrial sphere became the primary focus in the early stages of the industrial revolution with its host of new opportunities for investing in land space (Peters, 2010). Cities began to expand rapidly as populations grew and people flocked to cities for work. This shift in focus towards the terrestrial environment was to alter our perceptions of the oceans. For many in the twenty-first century, marine environments are on the edge of consciousness, largely invisible and seemingly irrelevant to our day-to-day lives. Most of us live on land and often out of sight of the sea. We perceive ourselves as a terrestrial people. Many of us, if we do happen to look at the sea, see an expanse of blue – not fully comprehending the three-dimensional environment that operates on and under the surface; even maps often show the oceans as empty blue featureless spaces. We have even also lost knowledge of where our food, including seafood, comes from. Yet, this disconnect is not true for all of us. Consider the role that the marine environment plays, for example, in the life of a whale biologist, a prawn fisher, or a sea kayaker. As pointed out by Shackeroff *et al.* (2009), the oceans are in fact 'peopled' seascapes; they affect and are affected by people.

Marine environments began to be conceptualised as what is known in geographical circles as 'spaces' – abstract locations with no social connection for humans, areas which are unknown and largely unexperienced. This is in stark contrast to terrestrial environments, which we understand as 'places' – areas which are created and valued by human experiences. The oceans and coasts are not just spaces, they are highly contested spaces. Space is an entity which is socially produced and reproduced; it is a frame of reference by which humans refer to the physical and social context.

Space is a meaningfully constructed social reality. Moreover, space is not static: space changes as planned uses, actual uses, and the meanings associated with those uses compete and evolve. Contestation arises when we hold differing social realities regarding the same space.

This is clearest in relation to the supposed polarisation between use and conservation of the oceans. In the same space, one group may want to install a wind farm, whereas others may want to establish an MPA. While more conservative political parties may back the push for development, more 'green' or environmentally focused parties may back the drive for protection.

Moreover, marine environments are contested spaces which are not flat nor fixed (like land), but rather voluminous and dynamic; they are spaces which are not just moved on, but also through and under. Oceans and coasts also have boundaries which can be fluid and fuzzy. Such properties make these contested understandings infinitely more complicated. In considering Steinberg and Peters' (2015) suggestion, we can think about sea spaces as a 'dynamic assemblage in which mobile human and non-human (including molecular) elements and affects are not merely passively consumed but imagined, encountered, and produced'.

Summary

The marine environment is substantially different from the terrestrial environment with many quite particular properties. The physical environment is truly three-dimensional rather than our two-dimensional perception of the terrestrial, it is fluid rather than static and borders aren't clear, and it is more influenced by temporal variability. It provides a wide range of ecosystem services including nutrient and carbon recycling, it provides flora and fauna for a wide variety of uses, and it is key to our health and wellbeing. These ecosystem services are considered 'common goods' in nature meaning that they are governed differently from many of those provided in the terrestrial environment. Lastly, we have a very different relationship with the marine environment than we do with the terrestrial. It is a largely 'unknown' environment with little social connection that is constantly conceptualised and re-conceptualised as we intrude further and further into its kingdom; it is a contested space. The marine and coastal environment is unique and such fundamental differences are likely to have ramifications for how we understand conflict in the two, quite distinct, spaces.

References

Berkes, F. 2005. Commons theory for marine resource management in a complex world. *Senri Ethnological Studies*, 67, 13–31.

Grotius, H. 1916. *Mare liberum*. Oxford University Press.

Hardin, G. 1968. The tragedy of the commons. *Science*, 162, 1243–8.

International Tanker Owners Pollution Federation Limited. *Weathering* [Online]. Available: www.itopf.org/knowledge-resources/documents-guides/fate-of-oil-spills/weathering/ [Accessed 15 May 2017].

Livengood, E. and Chapman, F. 2007. The ornamental fish trade: An introduction with perspectives for responsible aquarium fish ownership. *University of Florida IFAS Extension*, 1–8.

Longhurst, A., Sathyendranath, S., Platt, T., and Caverhill, C. 1995. An estimate of global primary production in the ocean from satellite radiometer data. *Journal of Plankton Research*, 17, 1245–71.

National Oceanic and Atmospheric Administration. *How much of the ocean have we explored?* [Online]. Available: https://oceanservice.noaa.gov/facts/exploration.html [Accessed 16 June 2017].

Nichols, W. J. 2014. *Blue mind: The surprising science that shows how being near, in, on, or under water can make you happier, healthier, more connected, and better at what you do*. Little, Brown.

Ostrom, E., Burger, J., Field, C. B., Norgaard, R. B., and Policansky, D. 1999. Revisiting the commons: Local lessons, global challenges. *Science*, 284, 278–82.

Peters, K. 2010. Future promises for contemporary social and cultural geographies of the sea. *Geography Compass*, 4, 1260–72.

Post, J. C. and Lundin, C. G. 1996. *Guidelines for integrated coastal zone management*. World Bank.

Reid, P. C., Fischer, A. C., Lewis-Brown, E., Meredith, M. P., Sparrow, M., Andersson, A. J., Antia, A., Bates, N. R., Bathmann, U., and Beaugrand, G. 2009. Impacts of the oceans on climate change. *Advances in Marine Biology*, 56, 1–150.

Sanchirico, J. N. and Wilen, J. E. 2007. Global marine fisheries resources: status and prospects. *International Journal of Global Environmental Issues*, 7, 106–18.

Sarukhán, J., Whyte, A., Hassan, R., Scholes, R., Ash, N., Carpenter, S., Pingali, P., Bennett, E., Zurek, M., and Chopra, K. 2005. *Millennium ecosystem assessment: Ecosystems and human well-being*. Island Press.

Shackeroff, J. M., Hazen, E. L., and Crowder, L. B. 2009. The oceans as peopled seascapes. *Ecosystem-based management for the oceans*, 33–54. Island Press.

Steinberg, P. and Peters, K. 2015. Wet ontologies, fluid spaces: Giving depth to volume through oceanic thinking. *Environment and Planning D: Society and Space*, 33, 247–64.

Thorne-Miller, B. 1999. *The living ocean: Understanding and protecting marine biodiversity*. Island Press.

Chapter 3

The fit of existing conflict theory

> Conflict [noun]
> 'an active disagreement between people with opposing opinions or principles'
> 'fighting between two or more groups of people or countries'
> (Cambridge English Dictionary)

Chapter 2 clarified the reasons why we need to understand the marine and coastal environment prior to considering conflict in such a setting: the contrasting physical space, the different natural resources, the distinctive governance considerations, and the particular way we perceive it. Yet further context is required. Before we begin to think specifically about marine and coastal conflict, we need to clarify the notion of conflict.

The definition above describes conflict as 'an active disagreement', and as such the term tends to be used in a variety of different ways. It often overlaps with other terms such as problem, issue, struggle, fight, social dilemma, dispute, debate, and controversy. These alternative terms tend to be used to differentiate between types of conflict (e.g. fight for violent conflicts and debate for conflicts of value). Conflict can also occur at a variety of scales (for examples, see Table 3.1), from the global level right down to a difference of opinion between two individuals (although conflict between two individuals is not of interest here, as we will see in the next section).

Much research has centred on the types of conflict described in the top two boxes; unsurprising given there have been 285 distinct armed conflicts identified since 1946 (Dupuy and Rustad, 2018). Such research tends to focus on conflicts between states or civil wars. Although many marine and coastal conflicts do occur between states, violent clashes are out-with the scope of this book. The Cold War has also been the target of research efforts, particularly by members of such groups as the Cold War History Research Centre at the University of Budapest, the Cold War Research Group at Charles University, and the Wilson Centre in Washington. This too, however, is out-with the scope of this book.

Table 3.1 Examples of different scales and types of conflict

	Large scale	*Small scale*
Violent	The Second World War was the most widespread global armed conflict known. It lasted from 1939 to 1945 and involved a majority of the world's nations. It is believed that a number of causes led to the war including anti-communism, expansionism, militarism, and racism. In terms of total deaths, the Second World War was the deadliest military conflict in history with over 60 million people killed.	The ethno-nationalist conflict in Northern Ireland (commonly known as The Troubles) is an example of a smaller-scale violent conflict. The conflict arose as a clash of identities between nationalists (those who identified as Irish) and the unionists (those who identified as British). Although the conflict has been between a small number of active participants, over 3,500 people have been killed as a result.
Non-violent	The Cold War can be viewed as an example of a large-scale non-violent conflict. It was a time (approximately between 1947 and 1991) of political and military tension between the United States and its NATO allies on one side and the Soviet Union on the other. It was a war between two superpowers fighting for dominance using psychological warfare, propaganda and espionage. Underlying the tension was the potential for nuclear war and global devastation.	Proposals for a third runway at Heathrow in the United Kingdom led to a great deal of non-violent conflict during the 1990s. On one side was BAA (those who proposed the expansion), the ruling political party of the time, and a number of influential British industries. On the other were the opposing political parties, local residents, and environmental groups. Actions arising from the conflict included rallies, flash mobs and marches as well as litigation.

As we will see, there are instances of violence in some of the conflicts that are described in this book. However, the focal point is conflicts which are much more common, insidious, and problematic for sustainable development. These are the small-scale, non-violent conflicts described in the fourth box.

Theorising conflict

To navigate our way to understanding why marine and coastal conflict occurs, we need to start with what we already know about conflict. It is useful to work through existing theory. A theory is just a way of explaining, of saying how things relate to each other, why they are the way that they are, and how they relate to other things.

The field of peace and conflict studies is extensive. The topic of conflict cuts across a range of disciplines, from economics to biology by way of history, political science, sociology, anthropology, and psychology. Theories dealing with classes of conflict, such as international conflict, strikes, race riots, guerrilla war, and ecological competition have received far more attention than any general theory that might be applicable to all manner of conflict. This means that it is only possible to touch upon a selection of conflict theories. In this chapter I explore how existing theories of conflict can help us to understand marine and coastal conflict. I review some of the work by early social conflict theorists, the works of later environmental (and natural resource) conflict theorists, and I assess if and how these theories address the peculiar characteristics of the ocean and coastal environment identified in the previous chapter.

Social conflict theory

'Social' conflict refers to a situation in which the parties are an aggregate of individuals, such as groups, organisations, communities, and crowds, rather than single individuals. One particular definition of social conflict conveys its meaning very well: 'social conflict [is] a struggle over values or claims to status, power, and scarce resources, in which the aims of the conflict groups are not only to gain the desired values, but also to neutralize, injure, or eliminate rivals'. (Coser, 1967: 232).

According to (Oberschall, 1978), social conflict theory attempts to explain three things:

1 The structural sources of social conflict, in particular structures of domination that make struggles over values and scarce resources likely.
2 Conflict-group formation and the mobilisation for collective action of challenging groups and their targets.

3 The dynamics of conflict such as processes of interaction between conflict groups; the forms of conflict; its magnitude scope, and duration; escalation and de-escalation; conflict regulation and resolution; the consequences of conflict outcomes for the contending groups and the larger society.

There is no one generalised theory of social conflict. However, various theories have emerged over time, and I provide a brief overview here. The history of social conflict theory can be traced back to Niccolo Machiavelli (1469–1527) and Thomas Hobbes (1588–1679), both of whom believed that conflict was a basic condition of humanity. Much current understanding of social conflict originated in the works of Karl Marx (1818–83), a German philosopher, sociologist, economist, and revolutionary socialist, and one of the most influential social thinkers to emerge in the nineteenth century.

In *The Communist Manifesto* (1967), Marx argued that capitalism would inevitably produce internal tensions which would eventually lead to its destruction. Experiencing early industrial Europe, he saw society as being made up of individuals, in different social classes, who must compete for social, material, and political resources such as food and housing, employment, education, and leisure time. He concluded that social institutions like government, education, and religion reflected this competition and helped to maintain the unequal social structure. This meant that some individuals and organisations were able to obtain and keep more resources than others, and these 'winners', in turn, used their power and influence to maintain social institutions. Furthermore, Marx considered that the capitalist system shaped how societies viewed people living in those societies (e.g. that the rich bourgeoisie had earned their wealth and the poor proletariat were lacking in skills and initiative) and that this created a class conflict. The system meant that the interests of the two were at odds and rcsources were unequally distributed among them.

It has recently been suggested that the problematic, often contradictory, relationship between humans and the rest of nature was a central theme in Marx's thinking throughout his life (Benton, 2018).

> Man lives on nature – means that nature is his body, with which he must remain in continuous interchange if he is not to die. That man's physical and spiritual life is linked to nature means simply that nature is linked to itself, for man is a part of nature.
>
> (Marx, 1974)

He supposed that people must meet their needs in harmony with one another and with the rest of nature, but that this relationship had gone wrong under capitalism.

Several social theorists built upon and refined Marx's original ideas regarding conflict and inequality. For example, German sociologist Max Weber (1864–1920) largely agreed with Marx but also thought that, in addition to economic inequalities, inequalities of political power and social structure also caused conflict. In *The Theory of Social and Economic Organisation* (2009) Weber noted that different groups were affected differently based on education, race, and gender, and that people's reactions to inequality were moderated by class differences and rates of social mobility, as well as by perceptions about the legitimacy of those in power. Where Marx concluded that the only struggle of importance was that between owners and workers, Weber argued that more than one conflict over property existed at any given moment in a given society and laid the groundwork for an understanding of pluralistic forms of social conflict in modern society (Runciman, 1978). He pointed out that societies are not always polarised into the opposing camps of the 'haves' and 'have-nots', but that every society is divided into various groupings with distinctive world views. Weber also recognised an emotional aspect to conflict; his conclusions like those reached by Sigmund Freud (1856–1939) and Friedrich Nietzsche (1844–1900).

A contemporary of Weber, Emile Durkheim, believed that solidarity was the normal condition of society, and even though he recognised the turmoil associated with industrialisation, he considered conflict to be abnormal or pathological. In *The Division of Labour in Society* (2018), he resolved the contradictions of the time by developing the notion of 'anomie', best understood as insufficient normative regulation. He argued that during periods of rapid social change, individuals sometimes experience alienation from group goals and values. They lose sight of their shared interests based on mutual dependence. In this condition, they are less constrained by group norms.

Believed to be the founder of modern conflict theory, American sociologist C Wright Mills (1916–62) built upon the work by Marx, Weber and others. In *The Power Elite* (1981), focusing on the United States, he argued that a only a few individuals within the political, military, and corporate realms held the majority of power within the United States and they made decisions that resounded throughout all American lives. He speculated that social structures were created through conflict between people with differing interests and resources and that individuals and resources were in turn influenced by these structures in addition to the unequal distribution of power and resources in society (ibid.). Essentially, the idea Mills proposed was that conflicts arise when resources, status, and power are unevenly distributed between groups; and society is structured to benefit a few at the expense of the majority.

Lewis Coser (1956), a contemporary of Mills, almost returned to the perspectives of Machiavelli and Hobbes, claiming that conflict must be recognised as a fundamental element of human relationships. However, Coser argued that the struggle behind conflict stimulates social innovation and change and he critiqued overtly stable social relationships, rigid social structures, and processes of centralisation for making dissent more likely. He reasoned that social conflict itself should be tolerated as naturally functioning to raise awareness of common norms, instigate conformity to laws, and, over time, to generate new social structures.

Such positive analysis of the role of conflict was extended by the next generation of sociologists, for example American sociologist Randall Collins (2015) elaborated upon the links between the macro-context and the micro-level of personal interaction and behaviour. He argued that each individual's constant struggle for status – and the conflict this entails – only makes sense by understanding the stratified environment in which that individual lives. Collins believed that while inequality and conflict predominate in modern society, competition and control are, in fact, the drivers of stability and progress within it. Along similar lines, David Lockwood, a British sociologist, argued in *Solidarity and Schism* (1992), that any theory of social order requires both a degree of integration and an element of class conflict.

Further development of sociological conflict theories has more recently emphasised the importance of detailed study of processes in society, moving away from the production of grand theory.

Environmental conflict theory

Environmental or natural resource conflict came to prominence on the conflict research agenda because, increasingly, environmental issues were assumed to play a rapidly rising role as the cause of violent conflict and threats to national security. Rather than building on the previous research by social conflict theorists, it arose as a new theme in international relations studies. Statements regarding the role of the environment in future international conflict had become ever more common, such as those by Galtung (1982: 99) 'destruction of the environment may lead to more wars over resources' and Opschoor (1989: 137) 'ecological stress and the consequences thereof may exacerbate tension within and between countries'. From the mid-1980s, scholars aimed at extending conventional security thinking to include other issues such as environmental change and resource depletion. Research aimed attention at whether, and under what circumstances, the biophysical environment represented a threat to national and global security.

In the early 1990s a number of major contributions appeared which concentrated upon empirically tracing the environment–conflict link. They were characterised by a strong emphasis on empirical evidence and the analysis of numerous case studies. This research stream focused predominantly on causal links between environmental scarcity, degradation, and acute national and international conflict in developing countries and countries in transition. Two research groups, led by Thomas Homer-Dixon in Canada (Toronto Group) and Guenter Baechler in Switzerland (Bern-Zurich Group) were at the forefront of this endeavour, and broadly confirmed each other's hypotheses that a positive link existed between environmental change and conflict.

The Toronto Group hypothesised that human pressure on natural resources could impact on a society's wellbeing and therefore contribute to the outbreak of violent conflict (violent conflict being the focal point of this group). As detailed in *Environment, Scarcity and Violence* (2010), this pressure was suggested to be environmental 'scarcity' comprising three main factors: degradation or depletion of resources (supply-induced scarcity), increased demand for resources due to population growth or increased consumption (demand-induced scarcity), and changes in access to resources due to unequal distribution of resources among groups (structural scarcity).

Homer-Dixon and his colleagues analysed a variety of conflict case studies from around the globe. Case studies included (among others): the Chiapas rebellion in Mexico where indigenous peasants were pushed into ecologically vulnerable highland and tropical forest areas; the role of rapid population growth and cropland scarcities in the Rwandan civil conflict; and conflicts between state-led resource extraction activities (particularly forestry) and local communities in Indonesia. The group proposed that scarcity influenced conflict indirectly by causing intervening social effects such as impoverishment, population displacement, or state weakening.

In *Environment, Scarcity and Violence*, Homer-Dixon further explained that environmental stress by itself does not cause violence. It must combine with other factors, such as the failure of economic institutions and government. He argued that the human ability to produce and deliver useful ideas (which he calls 'human ingenuity') can reduce the likelihood of conflict, particularly in countries with efficient markets, capable states, and an educated populace.

The Bern-Zurich Group produced similar findings but were rather more concerned with the negative consequences of degradation of the natural environment. They argued that human transformation of the environment frequently leads to 'environmental discrimination', i.e. that some actors, depending on their social, ethnic, religious, or regional identity experience inequality in their ability to access resources relative to others (Baechler, 2013). They perceived there to be three main ways in which humans caused environmental degradation: (i) overuse of renewable resources; (ii) overstrain of the environment's sink capacity (pollution); and (iii) impoverishment of the living space.

Like the Toronto group, this group's research also concentrated on violent conflicts (local and international), with examples mainly from the global South. Researchers from Bangladesh, Germany, Great Britain, Nigeria, and Switzerland carried out 40 area studies. Case studies included (among many others): the Palestine–Israeli water dispute and environmental degradation (desertification, deforestation) in the Horn of Africa. The group argued that environmentally caused conflicts escalate across the violence threshold only under certain conditions, and that human-induced environmental change can be either a contributing or a necessary factor to the emergence and/or the intensification of violent conflicts.

Based on their findings in the Environmental Conflicts Project (ENCOP), the group developed a typology of environmentally induced conflicts which included: (i) centre-periphery conflicts (urban vs. rural); (ii) ethno-political conflicts; (iii) inter-regional migration/displacement conflicts (induced by changes such as persistent drought, flood, and desertification); (iv) trans-boundary migration conflict; (v) demographically caused conflicts (high population pressure); (vi) international water/river basin conflicts; and (vii) international conflicts due to neo-colonialist exploitation of resources. This group theorised that the two key factors determining environmental conflict were discrimination against actors in sensitive ecological areas and a high level of dependence on natural capital. In *Violence Through Environmental Discrimination* (ibid.) Baechler went on to determine what causes such conflicts to become violent.

Researchers at the International Peace Research Institute in Oslo (PRIO) set out to test and validate or disprove these theories. They identified some missing links between environmental degradation and environmental conflict and published a nine-point critique of environment, population, and conflict literature, mostly criticising Homer-Dixon and his group. They suggested that there is a lack of clarity over what is meant by 'environmental conflict'; that researchers engage in definitional and polemical exercises rather than analysis; that important variables have been neglected, notably political and economic factors, which have a strong influence on conflict and mediate the influence of resources and environmental factors; that some models become so large and complex that they are virtually untestable; that cases are selected on values of the dependent variable; that the causality of the relationship is reversed; that postulated events in the future are cited as empirical evidence; that they fail to distinguish between foreign and domestic conflict; and that confusion reigns about the appropriate level of analysis (Gleditsch, 1998). They suggested that any analysis of resource use patterns really requires an understanding of the institutions that shape the rules and rights of resource use. Other scholars argued that the spotlight on the environment as a source of conflict prevented a more holistic view on the complex interactions between natural resources and human behaviour (Hagmann, 2005), and suggested that the relationship between groups and ecosystems is a product of social processes rather than of nature and should be viewed as a contested process that inscribes itself in the relations between different resource user groups.

Conflict is also at the core of many political ecology studies, a research approach which specifically engages with the causes and consequences of uneven power relations over natural resources and the environment. Core assumptions for researchers in this field include: the belief that nature and society are mutually interlinked; that it is not scarcity but rather entitlements, access, control, distribution as well as representation and meaning that lead to natural resource conflict; that the natural environment is an arena of contested entitlements and cultural meanings; and that environmental conflicts are social conflicts, are economic conflicts, are political, and are cultural conflicts. Clearly, this is a continuously developing area of research.

While the research into conflict, peace, and security has started to focus on environmental conflicts and the causes of those conflicts, this research tends towards large-scale, terrestrial, violent conflict. Conflict in ocean and coastal environments, however, mostly occurs at smaller, more local scales and tends to be less violent. That is not to say that it does not include key indicators of conflict such as organised parties, legal decisions, public attention, political disputes, and governmental action. However, although there are likely to be many similar causes of contention, the scale and level of violence, in addition to the fact that the marine environment is quite different to the terrestrial, may mean that differences also exist.

Finding the 'sea' in theory

As we have seen, there is a diverse range of theories into social conflict and environmental conflict. But, are any of these theories relevant to the problem of marine and coastal conflict? Can they deal with the unique characteristics of the ocean and coastal environment?

The function of the physical environment

Sadly, we have seen that there was little thinking by the social conflict theorists on the relations between humans and the natural environment. The predominant social conflict theories arose during the early industrial revolution, at a time when the terrestrial sphere became the primary focus for humankind. Social conflict theories were largely based around the concerns of the time. More recently, these theories have continued in much the same vein, concentrating mostly on societal structures such as class and elites. These theories do not specifically consider the physical environment in general, never mind the marine and coastal environment in particular.

In natural resource conflict theories, we see the physical environment most often referred to in respect of the form of natural resource over which the conflict is occurring. Environmental change is referred to (i.e. deforestation, desertification, water scarcity), but there is little consideration of how the physical environment itself influences the conflict. Of those natural resource theories which were grounded in empirical evidence, it does not appear that many marine and coastal case studies were included, instead focusing predominantly on the terrestrial sphere. Of those that were, several linked to fresh water conflict and its availability for human consumption.

It is conceivable that the nature of conflict becomes more complex when considered in three dimensions rather than two, multiplying the number of natural resources available within one single location. It is also credible that fluidity causes additional complexity in conflict, when pollutants can move faster and more widely than on land. Indeed, although not focusing on the lack of understanding of the physical environment in environmental conflict theories, in his nine-point critique Gleditsch (1998: 390) noted that:

> At sea, boundaries between states are not yet well settled, and even where they are established by law or by custom, they are not visible. The fluidity of any sea boundary makes it more conflict-prone than an established land boundary. Moreover, fish stocks straddle national boundaries and migrate across them with the seasons, with no concern for the consequences for human conflict. It is not surprising, then, that international fisheries should be ridden with conflict.

The role of environmental 'commons'

The 'commons' were recognised as relevant to social conflict theorists, particularly by Karl Marx. He argued that capitalism requires ongoing enclosures and appropriation of the commons for continuous accumulation and that this was realised through the dispossession of the commoners from the lands they had maintained and inhabited for decades, or even centuries. Indeed, Marx in *Capital: A Critique of Political Economy* (1976) gave vivid and harrowing accounts of the eviction of tenants from common land and of the appropriation of common land for private property by the landed aristocracy. Yet, while there are moves to privatise areas of the oceans, the marine environment is still largely governed as a global commons.

The environmental conflict theorists clearly shone a spotlight on the concept of ecosystem services, given the focus on conflicts around croplands, forests, and fresh water, even if not using the term explicitly. The issues of the commons, however, tend to be less well addressed. It has been argued that one of the most fundamental problems with the Toronto Group's approach is the acceptance of exclusion and enclosure as a given, leading to an understatement of the (often violent) processes that led to enclosure in the first place (Hartmann, 2003). This group, however, do refer to 'weak property rights' as a contextual factor in environmental conflict. For example, in a case study of the Philippines, Homer-Dixon (1994) notes that property rights are, for the most part, either non-existent or very unclear. He suggests that legally these areas are a public resource, and their 'open access' character encourages in-migration, yet many upland peasants find themselves under the authority of concessionaires and absentee landlords who have claimed the land.

The ENCOP project, led by the Bern-Zurich Group, considered the commons, at least from the perspective of governance. They suggested that conflicts over global commons are increasingly captured and mediated by a system of international regulations and conventions. They also noted that where conflicts do occur over the degradation or utilisation of regional commons, especially in the Third World, this tends to be characterised by a lack of regulatory mechanisms, regional institutions, norms, and contracts. However, they also argue that, in general, the progress made through environmental laws and conventions means that violent conflicts are unlikely to erupt in the foreseeable future (Spillmann, 1995). While this may be the case, this does not help us to further understand why conflict occurs in the global commons.

The nature of contested spaces

There is little consideration in conflict theory of how many areas, particularly those that are considered common, are contested spaces, and how this may influence conflict. Gleditsch (1998) pointed out that cultural factors are often missed in environmental conflict theories and this certainly appears to be the case in terms of how we understand and relate to either terrestrial and marine environments as places and spaces.

The Toronto Group recognises group identity as a contributor towards conflict, but these are discussed in terms of ethnic clashes, rather than how group identity relates to the spaces or resources over which the conflict occurs. The Bern-Zurich Group also discuss identity clashes – by tribes, clans, or ethnic groups – but from the perspective of how they are, in general, not willing to surrender land which they claim as the land of their ancestors. This is a nod to the contested nature of areas in which natural resource conflict occurs, but it is an aside rather than a central concern. Similarly, political ecology studies acknowledge that the natural environment is an arena of contested entitlements and cultural meanings. However, there are, as of yet, no general theories of environmental conflict arising from this new field of research, but instead a large number of empirical studies, making their theoretical contributions difficult to assess.

It seems practical that any theory of environmental conflict includes a consideration of the significance of the locational or spatial context around social phenomena. Indeed, scholars have identified the existence of place attachment, place identity, and place disruption when it comes to changing uses of the natural environment (Devine-Wright, 2009, Devine-Wright and Howes, 2010). While place attachment is understood as a product of emotional connection with a location, place identity refers to the construction of symbolic attributes to locations that create certain identities and attitudes. Alternatively, place disruption describes potential alterations to the physical and even social aspects and might also threaten pre-existing place attachments and identities, which may lead to specific behaviour or actions to cope with these disruptions. In highly contested spaces, such as the oceans and coasts, with multiple identities and attitudes, acknowledgement of the effects of place in any theory of conflict is imperative.

Summary

Conflict is a complex notion. It occurs on a spectrum from disagreement over individuals' values to a bloody, violent clash between groups. It also occurs at a range of scales, from between two individuals, to between several countries. For a long time, scholars have tried to understand what causes social conflict (i.e. the struggle for agency or power in society) and, more recently, natural resource conflict (the struggle for power over natural resources), with various theories proposed over the years. Such scholars have included social theorists such as Karl Marx, Max Weber, Emile Durkheim, and C. Wright Mills. They have also included natural resource conflict theorists such as Thomas Homer-Dixon, Gunther Baechler, and Nils Petter Gleditsch.

However, such existing theories of social and environmental conflict do not meaningfully engage with many of the peculiarities of the marine environment. There is little consideration of the physical environment, other than in the form of the resources it provides. There is some consideration of the 'commons' nature of marine and coastal resources and its requirements for different governing considerations. There is also little consideration of 'contested spaces' in the existing literature.

While this may not necessarily constitute a gap in how such theories assist our understanding of why marine and coastal conflict occurs, we must still consider it problematic. As such, we should consider the need for an understanding of conflict that addresses the oceans and coasts specifically.

References

Baechler, G. 2013. *Violence through environmental discrimination: Causes, Rwanda arena, and conflict model*. Springer Science & Business Media.

Benton, T. 2018. *What Karl Marx has to say about today's environmental problems* [Online]. Available: https://theconversation.com/what-karl-marx-has-to-say-about-todays-environmental-problems-97479 [Accessed 11 November 2018].

Collins, R. and Sanderson, S. K. 2015. *Conflict sociology: A sociological classic updated*. Routledge.
Coser, L. A. 1956. *The functions of social conflict*. Routledge.
Coser, L. A. 1967. *Continuities in the study of social conflict*. Free Press.
Devine-Wright, P. 2009. Rethinking NIMBYism: The role of place attachment and place identity in explaining place-protective action. *Journal of Community & Applied Social Psychology*, 19, 426–41.
Devine-Wright, P. and Howes, Y. 2010. Disruption to place attachment and the protection of restorative environments: A wind energy case study. *Journal of Environmental Psychology*, 30, 271–80.
Dupuy, K. and Rustad, S. A. 2018. Trends in armed conflict, 1946–2017. *Conflict Trends*, 5. PRIO.
Durkheim, E. 2018. The division of labor in society. In: *Inequality*. Grusky, D. Routledge.
Galtung, J. 1982. *Environment, development, and military activity: Towards alternative security doctrines*. Universitetsforlaget.
Gleditsch, N. P. 1998. Armed conflict and the environment: A critique of the literature. *Journal of Peace Research*, 35, 381–400.
Hagmann, T. 2005. Confronting the concept of environmentally induced conflict. *Peace, Conflict and Development*, 6, 1–22.
Hartmann, E. 2003. *Strategic scarcity: The origins and impact of environmental conflict ideas*. London School of Economics and Political Science.
Homer-Dixon, T. F. 1994. Environmental scarcities and violent conflict: Evidence from cases. *International Security*, 19, 5–40.
Homer-Dixon, T. F. 2010. *Environment, scarcity, and violence*. Princeton University Press.
Lockwood, D. 1992. *Solidarity and Schism: 'the Problem of Disorder' in Durkheimian and Marxist Sociology*. Clarendon Press.
Marx, K. 1974. Economic and philosophical manuscripts of 1844. *Moscow: State Publishing. House of Political Literature*, 42, 41–174.
Marx, K. 1976. *Capital: A critique of political economy*, 3 vols. Pelican Books.
Marx, K. and Engels, F. 1967. *The communist manifesto* [1848]. Trans. Samuel Moore. Penguin.
Mills, C. W. 1981. *The power elite* [1956]. Oxford University Press.
Oberschall, A. 1978. Theories of social conflict. *Annual Review of Sociology*, 4, 291–315.
Opschoor, J. B. 1989. North-South trade, resource degradation and economic security. *Bulletin of Peace Proposals*, 20, 135–42.
Runciman, W. G. 1978. *Max Weber: Selections in translation*. Cambridge University Press.
Spillmann, K. R. 1995. *Environmental crisis: Regional conflicts and ways of cooperation. Report of the International Conference at Monte Verità, Ascona, Switzerland, 3–7 October 1994*, Center for Security Studies and Conflict Research, Swiss Peace Foundation.
Weber, M. 2009. *The theory of social and economic organization*. Simon and Schuster.

Chapter 4

A collection of conflict cases

Marine and coastal conflict is pervasive. In early 2019, 2,673 cases had been reported to the Environmental Justice Atlas, which documents and catalogues environmental conflicts. Approximately one-quarter of these cases concerned issues in the ocean and coastal zone. Of those, 84 related to fish, 79 to sand and gravel extraction, and 32 cases to shrimp. Many more involved various types of renewable and non-renewable energy extraction, heavy metal extraction, and waste disposal. Many more conflicts are not found in the Atlas. This provides us huge potential to help us to understand why marine and coastal conflict occurs albeit that there is not always a great deal of available information on such conflicts.

In this chapter I present the synopses of 15 marine and coastal conflicts from around the globe (Figure 4.1). This provides some background to the chapters that follow, chapters which start to explain why conflict occurs in the marine and coastal environment. The case studies illustrate a broad

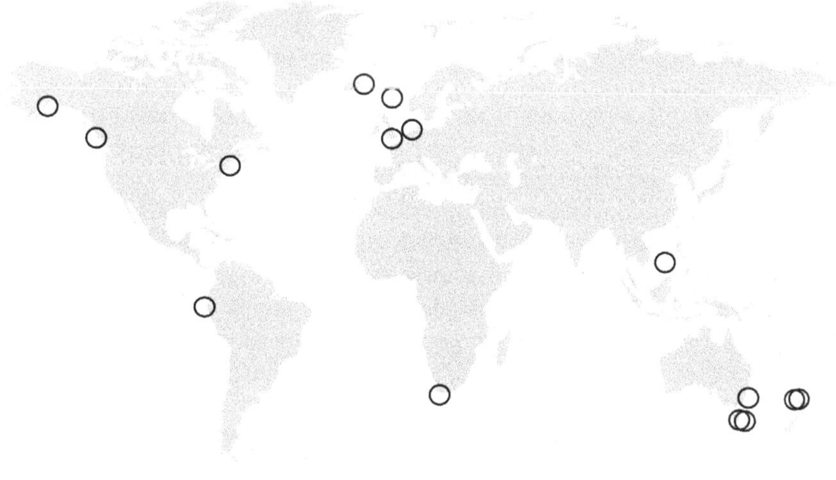

Figure 4.1 Case study locations.

range of marine and coastal resource conflicts regarding water, fish, oil and gas, wind and wave energy, mangroves, saltmarshes, and amenity.

Conflicts due to habitat destruction

Many areas of the world's oceans are experiencing habitat loss. Coastal areas, with their closeness to human population centres, have suffered disproportionately. These critical areas, which include estuaries, swamps, marshes, and wetlands, often serve as breeding grounds or nurseries for nearly all marine species. Yet, we dredge and fill in wetlands to accommodate urban, industrial, and agricultural development. It is estimated that every day between 1960 and 1995, 1 km of coastline was developed globally, causing permanent losses of valuable habitats, such as coastal wetlands, seagrass meadows, and rocky shores. This has caused conflict in developed and developing countries alike.

Muisne mangroves, Ecuador

In the 1970s, the Ecuadorian government began to encourage the adoption of shrimp mariculture, offering incentives such as low-interest loans for farm infrastructure. Rapid expansion caused Ecuador to become the number one producer by 1987, with most production in the Guaya and El Oro Provinces. The mid-1980s saw the establishment of intensive shrimp farming in the Muisne Canton. Shrimp ponds in Muisne estuary were owned by a number of commercial businesses, and came to occupy almost 85 per cent of the original area occupied by wetlands. Strong opposition came from locals and environmental associations, who defended the mangrove ecosystems in which the shrimp cultivation ponds were located.

Muisne, a town of 5,925 inhabitants (2010) is the political and economic centre of the Ecuadorian canton of the same name. It is located on an island near the coast on the south of the Esmareldas province. Mangrove ecosystems composed of *Acrostichum aureum*, *Avicennia germinans*, *Conocarpus erectus*, *Pelliciera rhizphorae*, *Rhizophora harrisonii*, and *Rhizophora mangle* border the island and mainland.

In this region, coastal communities have customary ownership of zones of mangrove. Until the early twentieth century, the higher parts of mangroves and the surrounding lands were virtually untouched. Later on, they were partly cleared for coconuts, bananas, plantains, cacao, and coffee. Communities recognised that personal work invested in agriculture created use rights over land plots. In contrast, the lower parts of the mangrove, a tidal area, were characterised as open access. Traditional livelihoods consisted of the selling and consumption of marine resources, mainly fish, shrimp, and the black mangrove cockles (*Anadara tuberculosa*). In addition, some people sell mangrove charcoal or cash crops such as cocoa and palm oil. When shrimp farming caused environmental degradation of the locality, this deprived the local population of food and income, and those living in the Muisne mangrove became involved in a large resistance movement against shrimp farmers.

In response to shrimp farming, the Foundation of Ecological Defence (FUNDECOL) was formed in 1989. Its aim was to claim access rights to lower mangroves for coastal populations as well as to preserve mangrove ecosystems for livelihood purposes. FUNDECOL mainly mobilised through legal actions and awareness raising through the media. One example is provided by Martinez-Alier (2001) who describes an action in July 1998 where Greenpeace linked up with FUNDECOL and other environmental groups in the sunrise destruction of one crop of shrimps. They opened a hole in a wall of an illegal pond and let the water flow out, then planted mangrove saplings. The presence of the crew of Greenpeace's Rainbow Warrior gave the necessary moral strength to the local groups, but the destruction of that illegal pond and the replanting were ideas proposed by FUNDECOL. Other, more cultural events were also regularly organised by opponents (e.g. cooking contests with mangrove products, wall paintings representing the fight against shrimp farming, festive events of marimba music, and theatre plays portraying the conflict).

The main grievances about the impacts of shrimp farming were environmental; particularly regarding mangrove destruction and shrimp farm discharge. These effects were believed to lead to a loss of traditional livelihoods, restricted access, marginalisation of classes, and other social conflict. The Environmental Justice Atlas highlights that other socio-economic impacts included an increase in corruption, displacement, a loss of traditional knowledge/practices/cultures, violations of human rights, and a loss of landscape/sense of place.

The provincial work done by FUNDECOL and its associated organisations has led to change. Examples include the formation of a national network, the Coordinating Committee for the Defence of Mangroves (C-CONDEM), a political platform for the movement; the creation of the National Environmentalist Assembly (Asamblea Nacional Ambientalista – ANA) a lobbying platform and a provider of policy advice; and the agreement by the government for a 500-hectare mangrove reserve.

Jade Bay dyke construction, Germany

Storm protection is essential for settlement in the coastal areas of Lower Saxony in north-west Germany. Since about AD 1,000, settlers there have co-operatively built and maintained coastal defence dykes, which today cover approximately 600 km of the coastline and protect more than 1.1 million inhabitants. There remains a strong tradition and awareness of the need for coastal protection efforts, particularly under a growing understanding about the importance of natural ecosystems. However, on many occasions, dykes have been enlarged and/or reinforced using land reclamation methods. Such methods diminish salt marshes.

In Jade Bay, the old dyke on the Cäciliengroden required upgrading in 1996. However, a national park had been established in 1985 ending at the dyke foot. The need to improve the integrity of the dyke led to conflict. Interventions in the dyke foreland and the salt marshes would be problematic. Conservationists led by Bund für Umwelt und Naturschutz Deutschland (German Federation for the Environment and Nature Conservation) obtained a temporary halt to initial dyke construction. They wanted more environmentally friendly construction methods to protect the sensitive salt marshes in the area. However, not everyone supported this halt.

Resistance against this decision began to stir when *Nordwest Zeitung*, a regional newspaper, published the headline 'Settlers get banners out again'. In the article they argued that salt marshes should not be more important than man. Community associations joined to distribute a leaflet calling for a 'human chain around Jade Bay'. This was supported by the Lord Mayor of Wilhelmshaven who called on the people to take part in the protest. On 30 August 1996, more than 10,000 people lit torches at dusk and spread out on the dyke between Cäciliengroden and Dangast; silently demonstrating for coastal protection, their own safety, and protesting the court verdict.

Aquaculture-related conflicts

Aquaculture is the farming of aquatic organisms in both coastal and inland areas. It is probably the fastest growing food-producing sector and now accounts for 50 per cent of the world's food fish (Kobayashi *et al.*, 2015). Aquaculture is often perceived to be a recent use of the marine and coastal environment. The modern era of commercial aquaculture, which came into prominence in the twentieth century, tends to use large-scale open-ocean finfish pen systems or shellfish rafts, and this is how we tend to think of mariculture. In the last few decades, worldwide aquaculture production has increased considerably, from 5 per cent of all fish production in 1970 to 32 per cent in 2000 (Tacon, 2003). This increase in production has coincided with an increase in related conflicts.

Okehampton Bay, Tasmania

Tasmania, an island State off the south coast of Australia has been the location for a series of protracted environmental battles, particularly over wilderness areas. An example from the late 1970s/early 1980s is the Franklin Dam project, a proposed dam on the Gordon River in Tasmania, Australia. It was never constructed. The campaign against the project, which eventually led to its cancellation, was one of the most significant environmental movements in Australian history. A history of complex and protracted conflict provides a unique context for this case study.

The Great Oyster Bay and Mercury Passage area is a relatively shallow bay bounded on the north and west by the Tasmanian coastline, and in the east by Freycinet Peninsula, Schouten Island, and Maria Island. Okehampton Bay is located on Tasmania's east coast, near the communities of Triabunna, Orford, and Spring Bay. There are two National Parks adjacent to the area, the Freycinet National Park and the Maria Island National Park. On the west coast of the northern section of Maria Island, there is a marine component to the park which prohibits commercial and recreational fishing and provides a sanctuary for a range of aquatic species. As in many other parts of rural Tasmania there has been a decline in the traditional activities of fishing, forestry, and agriculture. This has corresponded with a fall in employment opportunities, particularly for younger people, and a greater dependence on casual and part-time contracts, often from the increased reliance on tourism. The location has high recreational value and is a popular venue for sailing, boating, diving, surfing, and recreational fishing, particularly during summer when the local population is swollen with holiday makers. It is also a prime spot for holiday homes.

In 2017, Tassal, an Australian salmon farming company based in Tasmania, received approval to operate within a lease area, in Okehampton Bay, which was set aside for aquaculture purposes in the Great Oyster Bay and Mercury Passage Marine Farming Development Plan 1998 (subsequently reviewed in 2007 and 2010). The farm was to be located around 7 km from Spring Beach and more than 8 km from Maria Island, would have 28 salmon cages over 80 hectares, and was proposed as an experimental integrated multi-trophic aquaculture site, also growing mussels and seaweed. To support this, Tassal also planned to lease a purpose-built shore base from Spring Bay Seafoods. Independent economic modelling had indicated that the Okehampton Bay operation, once at full capacity, would provide an economic contribution to Tasmania of more than $80 million, and generate over 250 jobs throughout the economy (Tassal). Discourse around aquaculture in Tasmania focused on the way in which 'Big Salmon' transformed lives in depressed regional communities and acts as 'a perfect riposte to those mainlanders who paint the south isle as the "mendicant state"' (Denholm, 2016).

However, Tassal's proposed expansion provoked controversy and mobilised opposition from environmental non-governmental organisations (ENGOs), recreational fishers and sailors, tourism operators, and residents. In June 2017, protesters took part in 'FloatMo' (a pun on Dark Mofo – a festival held in Tasmania at the same time), a demonstration involving a variety of watercraft in Hobart's central Constitution Dock. The importance of tourism to the state's economy was often raised; this industry is argued to be a key part of the economy. Many protesters against aquaculture development in the state believed that it would damage the 'pristine' environment of its east coast, that Tasmania's 'brand' would be destroyed, and that it would hit tourism.

Northland Aquaculture Management Areas

The marine aquaculture industry in New Zealand has expanded from 2,800 tonnes in 1960 to 109,016 tonnes in 2016 (FAO, 2016). Government support is clear. Indeed, in 2012 the Government released their Aquaculture Strategy and Five-year Action Plan to Support Aquaculture stating their commitment to enabling the industry to achieve its $1 billion goal by 2025. Prior to 2004, aquaculture development was managed through a variety of local and centre government legislation, but deficiencies in this approach (such as lack of funding, lack of knowledge and lack of integration) led to a re-think. A key component of the subsequent reform was the establishment of Aquaculture Management Areas, which were to be defined within regional coastal plans (Banta and Gibbs, 2009).

The change in legislation simplified processes for applying for space for fish farms and accordingly promoted development of the industry. In the Northland area of New Zealand, this led to conflict with recreational sailors. The Northland Peninsula is in the far north of the North Island of New Zealand, stretching 330km from the Auckland isthmus. It has a convoluted coastline with many smaller branching peninsulas, making it popular for recreational sailors. Yachting New Zealand argued that 'the physical nature of aquaculture means that it will take up space in the marine coastal area that then cannot be used by others' (Sail World, 2011).

In one case, the Whaingaroa Fisheries Company, owned by Te Runanga o Whaingaroa proposed to build five hectares of sea cages for raising kingfish and a 19-hectare oyster farm next to Ohauroro/Peach Island. This was just inside the entrance to Whangaroa Harbour, approximately 30 nautical miles (nm) south of the Bay of Islands, and an ideal stopover for sailors on their way north. Opponents of the development argued that 'The farm will be just over 24ha in size and will virtually block the entrance to the bay' (Guest_TrackerMike, 2012).

Yachting New Zealand, who represented 31,000 members in 123 yacht clubs, working with Auckland Yachting and Boating and Northland Yachting Association, appealed against the Northland Regional Council Plan Change 4 (which related to aquaculture). They argued that the original plan did not take proper account of the need to consider and protect safe access to popular anchorages, harbour entrances, bays of refuge, navigational routes, and areas commonly used for yacht racing. As a result, several areas were established as off-limits to aquaculture development, including Whangaroa.

Broughton Archipelago

Similarly, the aquaculture industry in Canada has experienced substantial growth since 2000. In 2008, the Department for Fisheries and Oceans released Canada's Aquaculture Policy Framework, which committed to sustainable development of the industry as a key federal priority, a policy which is still ongoing ten years later. In 2015, British Columbia (BC) released a strategic growth plan to support several of BC's sectors, including aquaculture, to increase annual revenues to $15 billion per year by 2020. Aquaculture continues to be a priority area for growth in the region. However, this is not universally supported.

Although often referred to as the 'land question', the absence of treaties between settlers and the First Nations in British Columbia produces legal and political uncertainty for First Nations peoples regarding the oceans and coasts. Harris (2009) provides an overview of the complex history of the area in relation to native fishing rights, explaining that although BC's earliest settlers worked to maintain sovereign rights to marine resources, this was quickly eroded once BC became a Canadian province in 1871. Marine dispossession followed in the form of enclosure, privatisation, and the reallocation of resources. Few indigenous peoples were left with established rights to resources under treaty settlements.

Marine Harvest, a Norwegian-owned aquaculture business, operates within the traditional territories of 24 First Nations. The farms are sited in two main areas: (i) on Vancouver Island's Pacific North-West, around Tofino and (ii) on the east coast of Vancouver Island around the Broughton Archipelago. Marine Harvest holds formal agreements with 15 of the First Nations and seven First Nation owned businesses. At the centre of this conflict is Crown tenure renewal applications, which were due in June 2018, for several Marine Harvest farm sites in the Broughton Archipelago. This was further complicated by company restocking of fish farms in the area in 2017, against the wishes of First Nations and the provincial government.

A group of indigenous opponents, calling for the removal of fish farms from their traditional territories, held demonstrations near Marine Harvest farms on Swanson Island and Broughton Island. Protesters from the Musgamagw Dzawada'enuxw and Kwikwasut'inuxw Haxwa'mis First Nations began a peaceful, many months-long, occupation of two Marine Harvest fish farms on Swanson Island, setting up encampments and sleeping overnight on floating structures (Figure 4.2). Others participated in the occupation aboard a Sea Shepherd Conservation Society vessel. This ended when the Supreme Court ordered activists to remove their belongings and not to board or interfere with any of Marine Harvest's salmon farms operating in the area.

Figure 4.2 Indigenous families protest their defiance of Norway's Marine Harvest, the largest fish farm operator in the world in their traditional waters off the coast of Vancouver Island, Canada, at Swanson Island in May 2018.

Source: Mychaylo Prystupa.

Issues around this conflict related to traditional ownership rights; and the potential for altered quality of resources caused by farming-related diseases that infect wild fish populations and decimate the wild migrating salmon stocks upon which these groups rely for survival. These stocks are central to indigenous origin stories. Indeed, the Namgis origin story begins with a supernatural being who turns a man, Gwa'nalalis, into a river. 'There you will be a river for as long as the days dawn in the world, and you will be full of salmon so that your descendants may never starve', the being says (U'Mista Cultural Society, n.d.). The Kwakwaka'wakw people hold a Salmon Ceremony every year at the beginning of the salmon run. During the ceremony the head, bones, and entrails are separated from the flesh of the salmon. The fish is carried by twins, who are signs of good luck and believed to come from the Salmon People, to the river's edge where the chief thanks the Salmon People and returns the remains to the water. After the ceremony, the people gather to feast and perform a traditional salmon dance. However, some First Nations groups are concerned that salmon aquaculture will decimate the wild salmon stocks arguing that while the BC society at large and the economy in general would most likely survive the collapse or extinction of a run or stock of wild salmon, a First Nation whose distinct culture may depend on that stock, may cease to exist.

Macquarie Harbour

Macquarie Harbour is a large embayment (approximately 250 km^2) on the west coast of Tasmania, Australia. The Harbour has been the site of Atlantic salmon (*Salmo salar*) and Rainbow trout (*Oncorhynchus mykiss*) production for approximately 30 years. In 2011, Tasmania's three producer companies (Tassal, Huon Aquaculture, and Petuna) were producing 9,000 tonnes of fish in the area and had started working as a group towards doubling production by 2030. Federal approval was obtained for expansion, but initially capped at 15,000 tonnes (about half of what studies suggested the harbour could support). Using an adaptive management approach, the companies were to scale up slowly. In 2012, an expansion lease area was approved and subsequently a further increase in stocking led to production approaching 16,000 tonnes in 2015. By March 2016, the allowable cap for production was 21,500 tonnes, to be reviewed in early 2017. However, after the discovery of very low dissolved oxygen in the harbour and bacteria around farms in late 2016, the Environment Protection Authority (EPA) brought the stock cap back down to 14,000 tonnes (and further to 12,000 tonnes in 2017). The EPA also ordered the removal of fish from a lease which was owned by Tassal, showed signs of poor health, and was close to a World Heritage area. Furthermore, in May 2017, the EPA changed its allocation process from the equal 'tonnes per hectare' allocation to a 'percentage of current stock'.

These findings of negative environmental effects caused conflict within the industry, between the companies themselves, and from out with the industry. They came under severe criticism from concerned parties, particularly environmental groups. In 2016 and 2017, the effects of aquaculture attracted Australia-wide media attention due to a Four Corners (an investigative journalism program) report on pollution in Macquarie Harbour. The programme revealed that the EPA had found 'historic lows' of dissolved oxygen in the waters around fish farming sites in the Harbour, threatening endangered wildlife – raising public awareness of the issue and upping the conflict stakes.

Huon Aquaculture proceeded to take the government to court, with one case centred on whether the Tasmanian government and the EPA complied with federal conditions regarding salmon stock levels, and a second based on the technicalities of the EPA's decision to grant Tassal a higher biomass allocation.

Marine conservation-related conflicts

Considered an important instrument for conservation and for fisheries management, Marine Protected Areas (MPAs) are established to protect habitats, ecosystem structure and integrity, and species diversity and density. The benefits of MPAs are particularly evident in 'no-take' zones where the spill over of fish can profit local fisheries (Halpern *et al.*, 2009). Numbers of MPAs have increased around the world with global commitments, through the Aichi Biodiversity Targets, aiming for 10 per cent coverage of coastal and MPAs by 2020.

However, MPAs have also been criticised for leading to negative social, economic, cultural, and political impacts for local people and communities. In practice, they have led to convergent outcomes, from poverty reduction through tourism to inequitable sharing of economic benefits, without providing alternatives. This is problematic as support for, and the success of, MPAs is predicated on positive local perceptions of the outcomes.

Lyme Bay Conservation Area

Lyme Bay is an open stretch of southerly facing coastline in East Devon and West Dorset, extending 65 km along the English coast. Devon and Dorset are popular destinations and the tourist economy is an important income generator in the south-west due to its 'superb beaches and water-based activities' (VisitDorset, 2016). Dorset alone experienced 25.5 million day visitors in 2014, producing a total business turnover of £2.5 billion, and 13 per cent of all employment. There are also several coastal fishing towns adjacent to Lyme Bay, and fishing is a major employer, contributing 12 per cent of south-west GVA (Office for National Statistics, 2010).

In the early 1990s, local recreational divers expressed concern to the Devon Wildlife Trust (DWT) that scallop dredging was causing damage to reef communities in Lyme Bay. In 1991 and 1992, a survey carried out by the DWT identified that the marine life in the bay was of national and perhaps international nature conservation importance. It contained many fragile habitats such as maerl and eelgrass beds as well as a few rare species often at the extent of their geographical distribution range. The survey also confirmed that fishing had altered approximately 10 per cent of the reefs in the area. Traditionally, the fishers of Lyme Bay had towed demersal fishing gear (otter trawls, beam trawls, scallop dredges) on mixed sediment areas (sands and gravels) and placed pots in the rocky reef areas. However, by the 1990s mobile fishing gear was being used on the reef areas, resulting in the loss of static gear and causing physical damage to the reefs. Of concern was the reported impact of scallop fishing on the low-lying reefs which make up much of Lyme Bay, and especially on the pink sea fan, *Eunicella verrucosa*, which is a rare indigenous, slow growing, species.

Negotiations regarding how to protect reef features from fishing were held over a period of ten years. In 2001, the Southern Sea Fisheries Committee investigated the potential for a by-law to protect the reef features, but, facing opposition, they concluded that a voluntary approach would be more likely to succeed. An agreement was reached to protect two small areas of the Bay from bottom trawling and scallop dredging. However, scallop dredging was becoming more lucrative and an increasing number of boats travelled from other ports to fish in the bay, leading to the breakdown of the voluntary deal. In 2008, the Government took advice from its conservation advisors (Natural England) and enforced a statutory closure of 60 nm of reef as part of the Natura 2000 project, the largest closure of a marine area in British waters to protect wildlife. Increasingly larger areas of the bay were closed off to fishing between 2006 and 2010. Finally, in 2010, a wider area of 80 nm^2 was proposed as part of the Lyme Bay and Torbay Special Area of Conservation. Conflict broke out, largely through the media, between the fishing industry and environmentalists (specifically the DWT).

Environmentalists argued that destructive fishing practices, particularly scallop dredges, were destroying reefs and wildlife. They claimed that fishermen were turning the seabed into a desert and that using dredges to catch scallops was 'like picking mushrooms with a bulldozer' (Herald Express, 2007). Fishers pointed out that 'No-one, and certainly not fishermen, wish to destroy their stock, their ground or the marine environment, but the very nature of fishing demands that certain species are killed – killed so that we may eat' (Western Morning News, 2008). Fishing representatives argued that the voluntary agreement to avoid particular areas had been working and that the closure had a serious impact on the livelihoods of the fishermen who operated in the area.

Tsitsikamma National Park

In the Garden Route National Park, located on South Africa's southern coast, lies Tsitsikamma National Park. A coastal reserve covering a 50-mile long stretch of coastline, the Tsitsikamma MPA that lies within the park is the oldest in Africa, designated in 1964. Here, local communities have depended on natural resources as far back as the nineteenth century when woodcutters worked the forest and fished for subsistence, recreation, and income.

To protect inter-tidal and marine life, and to facilitate marine fisheries resource management by protecting spawning stock, the South Africa National Parks Board (SANParks) began to restrict fishing in the 1970s. First they cut shore-based angling to 15 sites along the western part of the park and then limited fishing in 1978 to a single 3 km stretch of the coast until 2001 when it was closed.

Throughout the period, the local community actively advocated to retain access to their historical fishing sites through the Tsitsikamma Angling Union. The fishers, alongside other recreational anglers who used the area wrote petitions, memorandums, and met with SANParks and other key officials. The situation was described as 'volatile'. In 2005, an organised protest march handed SANParks officials a memorandum demanding the restoration of fishing rights in the MPA. In 2006, an active campaign was launched by the community living next to the park, under the auspices of the Tsitsikamma Community Angling Forum, to restore their access and to open the park to a limited group. However, on advice from scientists, the Minister retained the park's no-take status in 2007.

After 20 years of discussions with fishers in the area, the Department of Environmental Affairs published draft regulations for the re-zoning of the MPA in 2015. Uproar followed, with criticism from the World Wildlife Foundation who said it was a 'parochial and political decision to appease highly vociferous local stakeholders' (Groenewald, 2007). The Orca Foundation set up a petition to save the marine reserve, for reasons as varied as lack of consultation, setting a precedent, the potential for long-term damage of fish stocks, and the loss of tourism business and other jobs. Furthermore, a group of concerned citizens, known as the 'Friends of Tsitsikamma' took legal action against the government department, SANParks and the Tsitsikamma Angling Forum after four designated areas of the National Park were opened to recreational shore fishing as part of a 'pilot project'. The Friends of Tsitsikamma won their case on 8 January 2016 after the authorities acknowledged a few breaches and a court order was made to stop fishing in the area immediately. Not long afterwards, Tsitsikamma fishers threatened violence against tourists and hikers to spur on the opening of areas of the MPA. Despite push-back from several conservationists, the Tsitsikamma National Park Marine Protected Area Declaration Notice was published in the Government Gazette 40510 while the regulations were published in the Government Gazette 40511 on Monday, 19 December 2016, allowing fishing in the park.

For conservationists, there were dire consequences of allowing fishing in the MPA, particularly the rapid decline of fish stocks, some of which were critically endangered, and the setting of a precedent which could compromise other protected zones. Others gauged it as a political issue,

> This is not about subsistence fishing. Most fishermen in the area already have other jobs. This is about destroying some of the last remaining fish stocks in the country, to gain votes at the next municipal election and keep politicians in power.
>
> (Ramsay, 2015)

Fishing in Tsitsikamma has been of both subsistence and cultural utility and research has found that a significant number of the community still want to fish in the park and, in fact, some have not ceased (Faasen and Watts, 2007). Many of the population consider themselves traditional fishers of Tsitsikamma and note that fishing forms a part of their culture and their local trading activities. One fisher pointed out:

> Our forebears fished in the area, and 50 years ago this tradition was ended without input from us. We are not planning to strip the coast of all its resources, we also need to protect it to ensure the next generation can continue the tradition.
>
> (Stander, 2015)

Fishers believed that the MPA was declared more than 50 years ago without public participation processes and that it stripped them of their cultural, historical, and recreational rights, that this was a product of inequity, discrimination, and social injustice.

Oil extraction-related conflicts

Offshore natural gas and oil production has a long history, with the first coastal oil rigs constructed in the late nineteenth century. One of the pioneers, industrialist Henry L. Williams, began extracting oil from the Summerland field off the Californian coast near Santa Barbara in the 1890s. Since then, offshore oil and gas extraction has become routine. Worldwide, around 92.6 million barrels of oil were produced daily in 2017.

The 'black gold' that is oil is understood to fuel violent conflict as authoritarian governments, as well as foreign powers and huge multinational companies, manoeuvre for control of oil fields. Oil also leads to conflicts of other types. Large-scale incidents, such as tanker disasters, account for 10 per cent of marine oil pollution in an average year, and, as the blowout at the Deepwater Horizon drilling rig in the Gulf of Mexico in 2010 showed, there can also be tragic consequences. Such incidents pose various conflict issues over the prevention, mitigation, and management of environmental disasters.

Brent Spar

By 1994, an oil storage buoy by the name of Brent Spar had been non-operational for five years. Its disposal was a problem for its owner Shell Expro (a subsidiary of Shell UK). The Brent Spar, a floating oil storage facility constructed in 1976 and moored in water approximately 140 metres deep, was in the Brent oilfield, 186 km north-east of the Shetland Islands, Scotland. Shell was not required by law to dispose of the buoy on land and indeed the International Maritime Organisation's guidelines stipulated that due to its size, it would be acceptable to sink it in the ocean. Shell commissioned 30 separate studies on the technical, safety, and environmental implications of its disposal and found that deep sea dumping within UK waters would be the best practicable environmental option, as it minimised damage to the environment, threats to workers' health and safety, and costs. Between December 1994 and the following May, the UK Department of Trade and Industry gave permission for Shell to sink the buoy in 2.5 km of water at the North Feni Ridge, about 160 miles off the west coast of Scotland. There followed a sustained, highly publicised campaign by Greenpeace, backed by European governments and the public.

On the face of it, the controversy over the Brent Spar disposal revolved around different perceptions of the likely impacts. The original impact assessment, which was produced for Shell by the consultancy AURIS, that the Spar would break up on hitting the seabed, possibly smothering organisms directly beneath it and in the near vicinity due to sediment disturbance. Additionally, the contaminants on board may leak or dissolve into the surrounding waters. The authors predicted, however, that the effects of this would be minimal. Greenpeace challenged this prediction of negligible harm on two grounds – that there was inadequate knowledge of the quantities of toxic materials on board the Spar; and that there was little understanding of the possible effects of such anthropogenic impacts on deep sea organisms and ecosystems. Both points were justifiable: the inventory of material on board was based on limited sampling, and the ecology of most deep sea organisms remained unstudied. However, many experts believed that the dumping would have had no greater impact than the sinking of any other ship at sea; that the environmental impact would be minute. Shell and the UK government vigorously defended the plan, claiming that Greenpeace was misinformed. Greenpeace, however, also saw this as an issue of precedence with concerns that it would become politically acceptable to dump disused offshore installations, of which more than a hundred could follow.

In protest, Greenpeace occupied the Brent Spar and so on 30 April 1995 Greenpeace's Moby Dick landed at the buoy and four climbers scaled up the rig. Activists then unrolled a banner reading 'Save the North Sea' (Figure 4.3). The controversy quickly hit the media headlines with pictures of Greenpeace activists braving water cannons from Shell's tugboats. Shell removed activists from the platform on 23 May, but the platform was again occupied by activists on 16 June.

Greenpeace gained support from politicians, in the UK and internationally. In June, the North Sea Protection Conference in Esbjerg, Denmark attracted Environmental Ministers from the countries surrounding the North Sea and, at the opening, virtually all the official delegates (except

Figure 4.3 Action at Brent Spar Oil Rig in the North Sea.
Source: Peter Thompson/Greenpeace.

the UK and Norway) condemned the sinking of the platform. In collaboration with other groups, a consumer boycott was requested of Shell gas stations, which was effective in Germany, Holland, and parts of Scandinavia. In many locations in Germany, sales slumped 20 per cent below average; 200 stations were threatened with attack; 50 were vandalised; two stations were firebombed; and shots were fired at another. Companies and public authorities also entered the fray by either cancelling contracts with Shell or threatening to do so.

Greenpeace, who had a positive image across Europe prior to the controversy, successfully used the media to gain public support, particularly through its occupation of the buoy. Activists conducted live interviews using satellite equipment and television relayed the North Sea events widely, with Shell portrayed as yet another greedy oil multinational behaving as they wanted.

On 20 June, after two months of protest, bad publicity, and growing public pressure, Shell backed down and agreed to dismantle and recycle the Brent Spar. Later, European nations at the OSPAR Convention agreed to a ban on the dumping of offshore steel oil rigs.

Exxon Valdez

At 12.04 a.m. on 24 March 1989, the tanker Exxon Valdez, rushing to leave the port of Valdez with just over 53 million US gallons of oil (produced onshore at Prudhoe Bay), ran aground on Bligh Reef. An 18-foot wide hole was ripped into the hull and nearly 11 million gallons of North Slope Crude Oil spilled into Alaskan waters at a rate of 200,000 gallons per minute. In the days that followed, the oil covered more than 2,100 km of coastline and 28,000 km of sea, hitting four National Wildlife Refuges, three National Parks, and Chugach National Forest. The immediate effects included the death of 100–250,000 seabirds, several thousand otters, harbour seals, bald eagles, killer whales, and an un-estimable number of salmon and herring.

In the clean-up process, the remaining oil was offloaded from the tanker within hours. However, the larger effort was hindered by an inadequate clean-up plan that had been created in the 1970s as well as by the nature of the location, situated in the Alaskan wilderness. Moreover, the various actors involved, including Exxon, the Alaskan State Government, and the US Government wasted time disputing the control of the situation, who would pay for what, who was responsible for what, and who would do the best job.

Several stakeholder groups were involved in the resultant social conflict – Exxon and the oil industry itself, the State of Alaska, and the administration of President George Bush, who had maintained that oil exploration could be undertaken in the region without environmental risk. Those most directly affected by the spill were Alaskan indigenous populations, commercial fishers, and landowners. For commercial fishers, the main grievance was the fact that pollution closed the fishery. Those able to fish were required to crowd into the same small cleaned-up area meaning dozens of near accidents and collisions. The Alaskan indigenous population saw the oil spill as a desecration of their economic lives but also their personal, psychological, social, cultural, communal, traditional, and religious forms of daily living. Even after the $2.5 billion clean-up, they were still afraid to eat the food from traditional hunting grounds. People in villages in the Sound depended on the waters near their homes for seal, sea lion, fish, and waterfowl. They hunted deer that browsed on the beachfronts. From the beaches, they harvested clams, mussels, molluscs, and other sea life. They claimed that a subsistence lifestyle thousands of years old had been irreparably damaged.

Hearings in the US Supreme Court and Court of Appeals stretched from 1994 to 2009 and resulted in punitive damages against Exxon of $507.5 million (reduced from the original $5 billion) plus $470 million interest for causing a public nuisance in a maritime setting. Commercial fishermen received compensation of $286.8 million and the Alaska natives settled for $20 million. Exxon had faced a boycott and widespread condemnation from environmental groups, backed by labour unions, marine pilots' associations, indigenous tribes, and fishing industry groups. The industry quickly lost legitimacy.

This catastrophic disaster also led to severe mental health problems among the communities of Prince William Sound, including anxiety, depression, and Post-Traumatic Stress Disorder arising from the direct impacts of the disaster, the effects of the clean-up (when crime rates soared and an unplanned increase in population meant that basic community needs could not be met), and the adversarial nature of the litigation.

Spratly Islands

The Spratly Islands, an archipelago located in the South China Sea, consists of over 100 scattered islands, shoals, banks, atolls, cays, and reefs spread across an area almost two times the size of the United Kingdom (410,000 km^2). The largest island spans less than 1.7 km and only 40 other features are considered islands. The remaining features of the archipelago are either submerged or only visible during low tide. It is mostly barren, uninhabitable, and contains little fresh water or other land resources, however it provides points from which claimant states can project jurisdiction over the water area and its resources.

The region holds potentially significant but largely unexplored reserves of oil and gas. Oil was discovered there in 1968 and was estimated by the Geology and Mineral Resources Ministry of the People's Republic of China (PRC) to hold reserves of oil and natural gas. It is a productive area for fishing. It is also one of the busiest areas of commercial shipping traffic in the world. Overall, it has economic and strategic importance and, consequently, is subject to an ongoing territorial dispute between China, Vietnam, Brunei, Taiwan, Malaysia, and the Philippines. This dispute has rumbled on from 1947 when China issued a crude map with 11 dashes, marking as Chinese almost the entire waterway.

While China, Taiwan, and Vietnam claim the whole archipelago primarily on historic rights; the Philippines, Malaysia, and Brunei want only certain islands or features, pushing their case on geographic proximity under the provisions of the United Nations Convention on the Law of the Sea. The littoral states have become entangled in a complex web of overlapping and conflicting claims over the sovereignty of the islands, and their associated natural resources.

For a while this conflict became violent. On 14 March 1988, shots were fired between Chinese and Vietnamese frigates after both countries dispatched troops to occupy Johnson South Reef in the Spratly Islands. The Chinese claimed that a landing craft was heavily damaged and finally sunk. The Vietnamese claimed that over 70 soldiers were killed in battle and two Vietnamese transport ships were sunk.

In 2016, a ruling by the Permanent Court of Arbitration, upholding the Law of the Sea Convention, nullified Chinese claims on historical rights. This means that if China continues to occupy reefs and develop them into military bases (a practice which has been happening since 2014), it will be acting outside of the bounds of international law. China rubbished the judgement and conflict rumbles on under the surface.

Renewable energy-related conflicts

The Kyoto Protocol, an international treaty extending the 1992 United Nations Framework Convention on Climate Change (UNFCCC) commits State Parties to reduce greenhouse gas emissions, particularly the most prevalent of those emissions, carbon dioxide (CO_2). With reducing oil and gas reserves as well as geopolitical uncertainty in many of the main production regions, concern has grown over energy security. This has led to government support, in many countries, for renewable energy technologies.

Solar and land-based wind energy technologies were an early focus. More recently, there has been a move to offshore wind (often to combat onshore siting concerns), and other marine renewable energy technology (particularly wave and tidal stream technologies). However, this is happening at a time of increasingly industrialised seas and competition for marine real estate oftentimes leads to conflict.

Cape Wind

Perhaps the best-known example of offshore wind conflict is the Cape Wind project in Nantucket Sound, US, which was hoped to produce one tenth of the electricity demand for the state of Massachusetts. The 130 turbines, covering 25 square miles, were to be placed in one of the most favourable wind energy sites on the east coast – at 'Horseshoe Shoal' in Nantucket Sound, a large body of water bordered by the southern beaches of Cape Cod and the islands of Martha's Vineyard and Nantucket. However, this led to a widespread, well organised, well financed, and politically potent campaign against the development.

Massachusetts actively promoted the development of a renewable energy industry with a target of 15 per cent of electricity output to be provided by renewables by 2020. Prices of electricity and natural gas were rising in New England and a stark reminder of the risks involved with imported energy occurred in 2003 when Cape Cod shell fisheries were closed after an oil barge spilled 100,000 gallons of oil into the area (Rodgers and Olmsted, 2008).

The Cape Wind scheme pitted environmentalist against environmentalist, resident against resident, and fishers against the developers. A myriad of actors inserted themselves, or were drawn, into the drama. Supporters of the project included the governor of Massachusetts and organisations such as Clean Power Now and Greenpeace, Union of Concerned Scientists, the Conservation Law Foundation, and the Cape & Islands Self Reliance. President Obama endorsed wind farms as one means by which to meet the nation's renewable energy and environmental goals. Multiple individuals and organisations opposed the project and the loudest voice among them belonged to the Alliance to Protect Nantucket Sound (Alliance), which was established by irate property owners, including the late Senator Edward Kennedy and 'fossil fuel magnate' Bill Koch, one of the Alliance's largest financial supporters. Well financed (reporting revenues of $1.74 million in 2010), the Alliance mounted an aggressive campaign against the project based on concerns such as migratory birds, tourism, visual impact, and threats to the commercial and recreational fishing industry.

Several fishing organisations took legal action against Cape Wind. The Martha's Vineyard and Dukes County Fishermen's Association launched a lawsuit against the government in 2010 claiming it could make navigation difficult, increase fishing costs and stop fishing in the site. This lawsuit was later dropped in 2012 after an agreement was formed to establish a 'permit bank' to enable the purchase of fishing permits. The Massachusetts Fisherman's Partnership (MFP), however, continue to oppose the project. In 2015, MFP, Inc and the Massachusetts Competitive Partnership filed briefs, both arguing Cape Wind had failed to meet the applicable Standard of Review and that an Extension Request should be denied. By 2017 the project was officially dead after developer Energy Management Inc pulled the plug on the controversial $2.6 billion project, surrendering its federal lease of 119 km^2 in Nantucket Sound.

Crest Energy, New Zealand

Approximately 75 per cent of New Zealand's (NZ's) electricity comes from renewable sources, particularly hydro, geothermal, and wind. NZ governments have also maintained an aspirational target of achieving 90 per cent renewable electricity generation by 2025. Marine energy is another way for the country to diversify its energy portfolio, and this is supported by the NZ government. Between 2008 and 2011, a government grant scheme, the Marine Energy Deployment Fund, was made available to support potential wave and tidal projects, with five projects supported, some of which have since been abandoned or have fallen dormant.

Kaipara harbour is one of the largest harbours in the world covering 900 km^2 with 3,000 km of shoreline. The Kaipara extends for 60 km north to south. The harbour has two factors that make it well suited to tidal energy generation – up to 8,000 million cubic metres of water pass in and out of the harbour each day leading to an average tidal range of 2.1 metres and commercial shipping rarely uses the harbour due to the treacherous tides and sand bars at its mouth. Crest Energy planned up to 200 completely submerged marine tidal turbines with a maximum capacity of around 200 MW, placed at least 30 metres deep along a 10 km stretch of the main channel.

The Crest Tidal Energy case underwent a lengthy approval process in New Zealand's Environmental Court, with the treatment of Māori cultural issues a defining feature. Although the significance of the area for Māori was recognised, the Court noted that Crest had 'adequately and appropriately' addressed the issues through 'extensive, considerable and meaningful' consultation and modification of its proposal. As a result, on 3 February 2011 the proposal to establish 200 'Open Hydro' style underwater tidal turbines at the mouth of Kaipara Harbour was consented in spite of Māori concerns.

Fishing-related conflicts

Fishing is a prehistoric practice that dates back at least 40,000 years although modern humans have been eating seafood for at least 164,000 years. Middens and cave paintings have proven the historic importance of seafood. By the twentieth century, coastal states were strongly supporting fishing, and a strong lobby developed to fight for traditional fishing rights.

Unfortunately overfishing (taking more fish from a body of water than can replenish itself) has stripped many fisheries around the world of their stocks. As much as almost half of the world's fisheries may be collapsed or overexploited (Branch *et al.*, 2011). Overfishing can mean that fish stocks crash suddenly or unexpectedly, or that their decline is gradual and predictable. Either way, this can lead to conflict.

Supertrawler Abel Tasman

The Australian small pelagic fishery extends from southern Queensland around to southern Western Australia. It consists primarily of sardine, blue and jack mackerel, and redbait. Stock assessments had concluded that the fishery was sustainable and not overfished, and that the total allowable catch (TAC) of 30,300 tonnes was conservative. It came from a strategy of taking less than 10 per cent of the assessed biomass.

In 2012, Seafish Tasmania, which held a quota of almost 18,000 tonnes, decided to expand operations in Australian waters by investing in a Dutch-owned factory freezer trawler named the F. V. Margiris, renamed the Abel Tasman on entry to Australia. Previously, the small pelagic fishery was fished from smaller vessels with no on-board processing or freezing capability. This restricted operations to close to land-based processing facilities otherwise the product could only be used as bait or in aquaculture feed. The advantage of the Abel Tasman's on-board processing facilities was that the operators could range widely throughout the fishery zone and the fish could still be processed for human consumption.

Having received approval from the Australian Fisheries Management Authority (AFMA), based on a scientifically devised quota, the plan was that on arrival in Australia the ship would immediately start harvesting the company's quota. If possible, the vessel would also harvest further quota leased from other operators. However, while the vessel was in transit from Europe, recreational fishers led a significant public debate, culminating in an unprecedented intervention from the Minister for the Environment.

Recreational fishers characterised the supertrawler as owned by a greedy multi-national company which would hoover up what little fish there were left in an attempt to make as much money as quickly as possible. 'We work hard to earn our annual leave and we won't stand idly by if politicians try and allow greedy multinational companies to decimate our vital marine resources and destroy the recreational fishing we all enjoy' (Hubbard, 2014).

This opposition resulted in the minister banning the vessel from fishing, amendments to national environmental laws, criticism of long-standing fisheries management practices, and two government reviews.

North Sea Mackerel Wars

Atlantic mackerel (*Scomber scombrus*) are widely distributed in the North Atlantic, extending from Morocco in the south to northern Norway as well as the Mediterranean and the Baltic. Mackerel spawn between February and July, after which they migrate into the Norwegian Sea and North Sea to feed. They tend to remain in these areas until August/September when they return to their overwintering grounds in the North Sea, west of the British Isles or in the Bay of Biscay. More recently, however, stock assessments indicated that the stock had expanded North-westwards during the summer feeding migration, potentially a consequence of changes in food availability, increased water temperature (increases of ~1–2 °C in the waters around Iceland), and/or increased stock size. The shift of mackerel stocks led to a substantial international conflict.

Until 2009, a tripartite agreement existed between the European Union, Norway, and the Faroe Islands regarding the TAC for north-east Atlantic mackerel. Each player received a fixed relative share of the yearly TAC. This was governed by several pieces of international law including the 1980 North-East Atlantic Fisheries Convention; the 1982 UN Convention on the Law of the Sea (UNCLOS); and the 1995 Straddling Fish Stocks Agreement. However, in 2009, Iceland was officially recognised as an additional player in the fishery due to the dramatic increase in the Icelandic catch of mackerel (from 38,000 tonnes in 2007 to 112,000 tonnes in 2008; Jensen *et al.*, 2015); catches in Icelandic waters constituted approximately 17 per cent of the total report landings.

In 2010, the EU and Norway established a ten-year bilateral pact on TAC allocations. Despite this agreement, a year after its banks collapsed, Iceland increased its catch by 23 per cent and the Faroe Islands increased its catch by 15 per cent, taking the harvest in 2010 above the recommended limits.

In response, Norway closed all its ports to Icelandic and Faroese trawlers in August 2010. Also, 50 pelagic fishers blockaded Peterhead port in Aberdeenshire, Scotland to prevent a Faroese vessel from offloading 1,100 tonnes of fish. The protesters used cars and vans to block a pump that carried fish from the boats docked in the harbour to a nearby processing plant. Later the Faroese claimed that police had failed to stop the incident which had caused a loss of £400,000, the catch having to go to meal and oil production and they warned Scotland that its excellent relations and bilateral trade agreements were at risk.

As the crisis escalated, the EU closed all its ports to Iceland and warned that their candidacy for full membership of the EU would be adversely affected. In 2013, the EU and Norway imposed additional trade restrictions on Iceland and the Faroes. In response, the Faroes threatened to take the EU and Norway to court through the United Nations.

The Mackerel Wars arose as a dispute between the EU and Norway on one side, and Iceland and the Faroes on the other, over the size of catches and quotas that countries could take from within their own territorial seas. Each country wanted what it considered was its 'fair share'. Scottish Fisheries Minister Richard Lochead accused Iceland and the Faroes of 'hoovering up' stock, taking more than their share. Scottish member of the European parliament (MEP), Struan Stevenson, suggested that that their behaviour was 'wholly irresponsible and bizarre' and that they were 'plundering stocks just like their Viking ancestors'. Iceland and the Faroes stated that they had been frozen out of talks with the EU and Norway and that any negotiations were all about preserving EU and Norwegian shares of the resource. Tomas Heidar, Iceland's chief negotiator said:

> The EU and Norway are not the sole owners of the mackerel stock and by taking almost all the recommended total allowable catch, they disregard the legitimate rights of the other two coastal states Iceland and the Faroe Islands, as well as Russia.
>
> (Sverisson, 2011)

Summary

In this chapter I presented synopses of 15 marine and coastal conflicts from around the globe, although mostly from North America, Europe, and Oceania. These case studies were split up according to the types of conflict that they represented: habitat destruction, aquaculture, conservation, oil extraction, renewable energy, and fishing. The conflicts occurred at a variety of scales, from the international to the local. The conflicts also used myriad methods by which to mobilise the conflict, such as media campaigns, lawsuits, and vigilante means. This overview has provided a real sense of just how diverse marine and coastal conflict can be. On the surface, these conflicts appear to be about geography, about how we relate to the resources within the geography, historically, currently, and into the future. They also appear to be about a crisis in democracy – we no longer trust that our interests will be represented. But what lies beneath the surface? What motivations are hidden in the deep?

References

Banta, W. and Gibbs, M. 2009. Factors controlling the development of the aquaculture industry in New Zealand: Legislative reform and social carrying capacity. *Coastal Management*, 37, 170–96.

Branch, T. A., Jensen, O. P., Ricard, D., Ye, Y., and Hilborn, R. 2011. Contrasting global trends in marine fishery status obtained from catches and from stock assessments. *Conservation Biology*, 25, 777–86.

Denholm, M. 2016. Dangerous dinners: Is salmon farming ruining Tasmania? *The Australian*, 22 December.

Faasen, H. and Watts, S. 2007. Local community reaction to the 'no-take' policy on fishing in the Tsitsikamma National Park, South Africa. *Ecological Economics*, 64, 36–46.

FAO. 2016. Aquaculture production (metric tons). Available: https://data.worldbank.org/indicator/ER.FSH.AQUA.MT/ [Accessed 5 March 2019].

Groenewald, Y. 2007. Fishing for a fight over Tsitsikamma. *Mail and Guardian*, 30 August.

Guest_TrackerMike. 2012. *New marine farm in Whangaroa Harbour* [Online]. Crew.org.nz. Available: http://crew.org.nz/forum/index.php/topic/6898-new-marine-farm-in-whangaroa-harbour/ [Accessed 28 November 2017].

Halpern, B. S., Lester, S. E., and Kellner, J. B. 2009. Spillover from marine reserves and the replenishment of fished stocks. *Environmental Conservation*, 36, 268–76.

Harris, D. C. 2009. *Landing native fisheries: Indian reserves and fishing rights in British Columbia, 1849–1925*. UBC Press.

Herald Express. 2007. Call for ban on scallop dredges. *Herald Express*, 27 November.

Hubbard, R. 2014. *Fishers and conservationists seek election commitment to permanently ban super trawlers* [Online]. Available: www.stopthetrawler.net/election_campaign_launch [Accessed 11 January 2018].

Jensen, F., Frost, H., Thøgersen, T., Andersen, P., and Andersen, J. L. 2015. Game theory and fish wars: The case of the Northeast Atlantic mackerel fishery. *Fisheries Research*, 172, 7–16.

Kobayashi, M., Msangi, S., Batka, M., Vannuccini, S., Dey, M. M., and Anderson, J. L. 2015. Fish to 2030: The role and opportunity for aquaculture. *Aquaculture Economics & Management*, 19, 282–300.

Martinez-Alier, J. 2001. Ecological conflicts and valuation: Mangroves versus shrimps in the late 1990s. *Environment and Planning C: Government and Policy*, 19, 713–28.

Office for National Statistics. 2010. *Portrait of the South West* [Online]. Available: www.ons.gov.uk/ons/rel/regional-trends/regional.../portrait-of-the-southwest.pdf [Accessed 1 December 2017].

Ramsay, S. 2015. *Tsitsikamma MPA pilot: Destroying 50 years of conservation?* [Online]. Available: www.traveller24.com/Explore/tsitsikamma-mpa-pilot-destroying-50-years-of-conservation-20151215 [Accessed 19 March 2017].

Rodgers, M. and Olmsted, C. 2008. The cape wind project in context. *Leadership and Management in Engineering*, 8, 102–12.

Sail World. 2011. *Yachting New Zealand makes stand on aquaculture intrusion on cruisers* [Online]. Available: www.sail-world.com/Australia/Yachting-New-Zealand-makes-stand-on-Aquaculture-intrusion-on-cruisers/-90783?source=google [Accessed 28 November 2017].

Stander, Y. 2015. Lifting fish ban fury. *The Times*, 30 November.

Sverisson, S. 2011. It's wrong to blame Iceland for overfishing mackerel. *Guardian*, 18 August.

Tacon, A. J. 2003. Aquaculture production trends analysis. In: *Review of the state of world aquaculture*. FAO.

Tassal. 2017. Tassal's Okehampton Bay salmon farming operation. Available: https://tassalgroup.com.au/okehampton-bay-salmon-farming-operation/ [Accessed 5 March 2019].

U'Mista Cultural Society. n.d. *Gwa'Nalalis: The man who became the river Gwa'ni* [Online]. Available: https://umistapotlatch.ca/enseignants-education/cours_1_partie_2-lesson_1_part_2-eng.php [Accessed 30 November 2017].

Visitdorset. 2016. *Dorset Tourism Data* [Online]. Available: www.visit-dorset.com/trade/research/dorset-tourism-data [Accessed 1 December 2017].

Western Morning News. 2008. Marine conservation must be team effort. *Western Morning News*, 4 November.

Chapter 5

The changing marine social-ecological environment

As we saw in the previous chapter, conflict occurs across myriad marine and coastal settings, often linked to the status of the marine environment as one of our most valuable resources. It is a source of food; used for transportation; mined for oil, gas, and minerals; and provides a limitless source of tourism and recreation opportunities. Although humans have been capable of travelling by sea and of fishing since Paleolithic times (40–50,000 years ago) (O'Connor *et al.*, 2011), our move into the Anthropocene era has meant that human activities in and near the world's oceans have increased exponentially.

There are countless ways in which marine and coastal resource use has increased. For example, global fish production (wild-capture and fisheries) has increased from 127.8 million tonnes in 2002 to 175.2 million tonnes in 2017 (Statista, 2019b). The capacity of container ships was almost ten times higher in 2016 than it was in 1990 (Statista, 2019a). In 2007, global production of oil was just over 82 million barrels, and by 2017 this was nearly 93 million barrels (BP Energy Economics, 2018), with offshore oil production contributing nearly 30 per cent (U.S. Energy Information Administration, 2016). In 2009, cruise tourism attracted 17.8 million passengers, often to remote and fragile areas such as the Arctic and Antarctic regions, with a forecast of 27.2 million in 2018 (Cruise Lines International Association, 2018). Various artificial islands have also been created, the largest being the Palm Jumeirah (covering 560 hectares) in Dubai. Furthermore, the human race appears to be committed to finding more and more ways to extract minerals such as silver, gold, copper, manganese, cobalt, and zinc and to plumb the oceans' depths for new medications to treat cancer, pain, and other ailments.

This expanding use of the oceans and coasts is creating a changing social-ecological environment and is driven by three key factors: a booming global population; its need for water, food, and energy security; and the imperative for ongoing economic growth. The latter two drivers were evident in the conflict case study analysis. At the same time, we have experienced increasing environmental degradation caused by overfishing, habitat loss, marine debris, invasive species, and more. Illegal, unregulated, and unreported fishing affects approximately 20 per cent of the global fish yields, which cost around US$23 billion a year; about 20 per cent of the world's coral reefs have been destroyed and show no immediate prospects for recovery; and plastic waste kills up to 1 million sea birds, 100,000 sea mammals, marine turtles, and countless fish each year (United Nations, 2017). This suggests that we are unable to manage human impacts on the oceans and coasts. Indeed, a failure of governance discourse also emerged in the case study analysis. When considering this changing marine environment and the effect it has on conflict, we must also include climate change. These five drivers of current (and potentially future) marine and coastal conflict will be the focus of this chapter.

Increasing population

In the early twenty-first century, we are in a time of overpopulation. That is, a time when the number of currently living humans exceeds the maximum number of humans that the environment can sustain indefinitely, given the food, habitat, water, and other necessities available. The global population grew from 1 billion in 1800 to 7.616 billion in 2018, continues to grow at around 83 million, or 1.1 per cent, per year and is projected to reach 9.6 billion by 2050 (Gerland *et al.*, 2014). The question of how many people the Earth can support is a long-standing one that becomes more critical as the population, and our use of natural resources, keeps booming.

There are many reasons for this overpopulation. Since the Industrial Revolution in the late eighteenth century, standards of living have risen, leading to a decline in the death rate. Technological advancement has improved food production. Advances in livestock production, from climate control to the nutritional qualities of feed, have boosted agricultural output. Refrigeration and modern packaging technologies increase the safety of our food and the distance across which it can be transported. Food biotechnology has enhanced the beneficial traits of plants, animals, and subsequently their food components.

There have also been discoveries to defeat a whole litany of diseases. Illnesses that had claimed thousands of lives are now prevented because of the invention of vaccines. The advent of antibiotics hailed a new era in the treatment of communicable disease. Ventilators, pacemakers, and other machines have been developed to support, enhance, or replace the body's organs, and the ability to transplant organs between individuals tied together surgery and immunology.

For thousands of years, those living in poverty had more offspring to counteract the high infant mortality rate, a pattern which has continued in areas still in poverty or that have suffered from natural disasters. However, compared to earlier times, many of these children now survive. In many of the developing nations where we still see bigger families, many people are illiterate and have little or no knowledge about family planning. For example, from 1950 to 2010, Sub-Saharan African population grew over three and a half times, from about 186 million to 856 million, whereas European population only increased by 35 per cent (Cleland, 2013).

With this growth has come a commensurate degradation of the environment and depletion of natural resources. Finite natural resources, such as fresh water, arable land, and frontier forests, continue to plummet. In terms of the marine and coastal environment, we know that close to 90 per cent of the world's marine fish stocks are fully exploited, overexploited, or depleted, and scientific estimates have shown that 64 per cent of the world's wetlands have disappeared since 1900 (Ramsar, 2017). The reduction in natural resources is placing stress on basic life sustaining resources. Meanwhile the overuse of non-renewable energy resources such as coal, oil, and natural gas, has affected the quality of air and has led to a rise in the amount of CO_2 emissions and a changing climate.

Population growth affects ocean and coastal health in several ways. Coastal skylines are rising ever higher with the construction of homes, hotels and restaurants, coastal transport infrastructure (e.g. roads, railways, shipping canals), and tourism and recreational infrastructure (e.g. marinas, boat ramps, and swimming enclosures). Entire cities rise from the ocean, the latest of which is the Chinese-developed Forest City project which aims to develop, by 2045, four artificial islands in Malaysia covering 14 km^2 of ocean (Country Garden Malaysia, 2015). Many of these developments involve land reclamation and must be defended from the seas by e.g. seawalls and breakwaters. However, such development causes large changes to the local environment through direct impacts (e.g. land filling, dredging, coral and sand mining for construction) and indirect impacts (e.g. increased run-off of sediment and pollutants). These can lead to smothering, or increased disease and mortality of ocean and coastal flora and fauna.

Population growth has also contributed to climate change and warming oceans. A 2009 study of the relationship between population growth and climate change found that the 'carbon legacy' of just one child can produce up to 20 times more greenhouse gas than any one person can save (Murtaugh and Schlax, 2009). The impacts of anthropogenic climate change on the oceans, as identified so far, include: decreased ocean productivity, altered food web dynamics, reduced abundance of habitat-forming species, shifting species distributions, and a greater incidence of disease (Hoegh-Guldberg and Bruno, 2010).

And, of course, an increasing population requires more resources. Satisfying the expected food demand will require a substantial increase of global food production of 70 per cent by 2050 (FAO, 2009). In their 2017

Energy Outlook, BP projected that although rapid gains will be made in energy efficiency, energy demands will still increase by 30 per cent by 2050 (BP, 2017). The World Wildlife Fund reported in 2006 that resource demand exceeded supply by approximately 25 per cent, meaning the earth took 15 months to produce the resources that had been consumed over 12 (WWF, 2016). In 2007, physicist Martin Desvaux argued that human impact on the planet has exceeded sustainability since the 1980s and that a sustainable global population is between 2–3 billion people (Desvaux, 2007).

With about 40 per cent of the world's population living within 100 km of the coast, it is not surprising that our activities are starting to take their toll. United Nations' figures predict that this may increase to 75 per cent within the next century (UN Oceans, 2008). As we start to run out of available land resources, we are increasingly casting our sights out to sea.

Security of supply

Concerns around the ongoing availability of the resources required to sustain the rising population means that the need for water, food, and energy security has become an imperative. Water security is defined as access to safe drinking water and sanitation. Energy security refers to access to clean, reliable, and affordable energy services for cooking and heating, lighting, communications, and productive uses. Food security means that we can access sufficient, safe, and nutritious food to meet our dietary needs and food preferences for an active and healthy life. The interactions between each are numerous and substantial. For example, many forms of energy and food production are water intensive; and energy is needed for extracting, transporting, distributing, and treating water as well as fuelling land preparation, fertiliser production, irrigation and the sowing, harvesting and transportation of crops.

While water security was not directly referred to in the case studies analysed, it is essential to include a summary of the issues due to the interlinked nature of the water-food-energy nexus. Water scarcity already affects every continent. Water use, mostly for food and fuel, has been growing globally at more than twice the rate of population increases over the last century. By 2025, it is expected that 1.8 billion people will be living in countries or regions with absolute water scarcity, while two-thirds of the over population could be under water stress conditions (Swyngedouw, 2013). In response, the world is now looking to the oceans for fresh water. Over the last 30 years, desalination (a process that extracts mineral components from saline water) has evolved into a viable alternative water supply. The sea is now viewed as the unlimited source from which we can create fresh water, with over 16,000 desalination plants worldwide in 150 countries. Although much of the ongoing conflicts over water security focus on freshwater, small-scale conflicts have arisen around desalination plants. For example, concerns were raised that not enough environmental studies had been conducted prior to the establishment of the Wonthaggi desalination plant in Victoria, Australia.

Unlike the newer technology of desalination, offshore oil and gas extraction has been around since the late 1800s when the first submerged oil wells in saltwater were drilled in the Santa Barbara Channel in California. In the early twenty-first century, however, offshore oil and gas extraction began to drive conflicts around the globe. Oil and gas are perceived as central to disputes over maritime boundaries between Russia and Norway in the Barents Sea and Australia and Timor-Leste in the Timor Sea. This is particularly clear in the maritime dispute over the Spratly Islands in the South China sea:

> China, which imports more than half its oil, will nearly double its demand for it over the next quarter-century, according to the International Energy Agency in Paris. Its demand for natural gas – believed to be particularly abundant beneath an archipelago of contested islands and reefs known as the Spratlys, just west of here – is projected to more than quadruple.
>
> (Higgins, 2011)

In the Exon Valdez oil spill case, it was also suggested that:

> The stalemate that ensued was resolved in 1973 when then President Richard Nixon and the U.S. Congress, reacting to the possibility of fuel shortages in this country and to the repeated assurances of oil industry and government experts that policies and procedures would be put in place to meet environmental contingencies, cut off the court challenges that had delayed construction of the pipeline.
>
> (Bowen and Power, 1993: 99)

Concurrently, there have been developments in the sphere of ocean energy. The potential for wave and tidal energies is huge, although the technologies are still largely undeveloped with few commercial projects in place. Offshore wind, however, is the most advanced technology. At the end of 2017, the total worldwide offshore wind power capacity was 18,814MW (Global Wind Energy Council, 2017) with all of the largest offshore wind farms in Northern Europe. Energy security oftentimes is used as an argument for the deployment of offshore renewable energy installations. For example, in the Cape Wind conflict, it was suggested that 'the wind farm would enhance energy independence while providing lower energy costs, a healthier environment and new jobs' (Miga, 2006). It was also suggested 'The administration has pushed a "green energy" agenda nationally as a way to create jobs and lessen U.S. dependence on oil imports.' (Sperry, 2012). The need for alternative energy sources was also raised in the Crest Energy conflict 'Government has been so keen to find alternative energy sources, it has backed technology that is untested' (Radio New Zealand, 2011).

The need for enough food is also a driver of conflict. The 1996 World Food Summit Plan of Action states that 'Food security exists when all people, at all times, have physical and economic access to sufficient, safe and nutritious food to meet their dietary needs and food preferences for an active and healthy life' (FAO, 1996). In 2016 it was believed that the number of chronically undernourished people in the world was around 815 million (FAO *et al.*, 2017). A famine was declared in South Sudan in 2017 and alerts for high risk of famine were issued for north-east Nigeria, Somalia, and Yemen. Fish is a critical food source in addressing many such famines, and capture fisheries and aquaculture provide some 3 billion people with almost 20 per cent of their per capita intake of animal protein and a further 1.3 billion with about 15 per cent (HLPE, 2014). Fishing is also a means of subsistence for billions of people. Subsistence fishing is when fish are caught to feed the family or the community of the fisher. Examples might include inter-tidal food-gatherers on sandy and rocky shores, shore-based lift nets, or traditional trap fisheries. This has meant that where areas have been closed to fishing for conservation purposes, or for the development of another industry, the need for food security has been used as an argument against closure. For example, in the Ecuadorian conflict over shrimp aquaculture, it was argued that restricted access and habitat loss was resulting in the decline in fisheries and creating food insecurity 'One housewife emphasizes that the decline in fish is also detrimental for the locals because this resource is one of their major food sources' (Crider, 2012: 43). In the Tsitsikamma conflict in South Africa, discussions around the protected area closure noted that conservation of natural resources took a fortress conservation approach, seeking to preserve natural resources in such a way as to coercively exclude 'rural communities that traditionally depend on these resources' (Faasen and Watts, 2007).

As capture fisheries have declined worldwide, we have seen an increase in mariculture (the farming of marine fish, shellfish, and algae). Aquaculture provided only 7 per cent of fish for human consumption in 1974, but this rose to 26 per cent in 1994 and 39 per cent in 2004 (Subasinghe, 2017). By 2030 it will be an estimated 62 per cent (Kobayashi *et al.*, 2015). Food security, it has been suggested, has driven the growth of aquaculture. Food security is also utilised as a reason for aquaculture development projects by their proponents. In relation to the conflict over aquaculture in Whangaroa Bay in New Zealand, one source noted that 'demand for aquaculture products is expected to strengthen significantly as the world's population grows and wild-catch levels remain relatively static' (Aquaculture New Zealand, 2011). In the aquaculture conflict in British

Columbia, 'Salmon aquaculture has become important as a major source of protein' (Gerwing and McDaniels, 2006: 260). These are less likely to be arguments for food security per se, but more for ensuring the burgeoning middle class has its favourite salmon products.

The economic imperative

At the 2012 United Nations Conference on Sustainable Development (Rio+20) marine issues and questions of oceans' governance became more formally negotiated than ever previously. This conference led to the term blue economy (interchangeably, blue growth/wealth, and ocean economy) becoming common parlance in marine governance. This was based on increasingly high state and private interest in oceans' development and conservation. The term was consistently used to draw attention to oceans and connect them to Rio+20's green economy (an economy that aims for sustainable development without degrading the environment) theme. In a keynote address at the conference, announcing the World Bank's plans for its 'Global Partnership for the Oceans', then World Bank President Robert Zoellick stated that the:

> Oceans are the home of an under-recognized and under-appreciated 'blue economy'. At a time when the world is looking for sources of growth, there is huge potential for 'blue growth' – wisely preserving and investing in the value of ocean ecosystems to fight poverty and improve lives.
>
> (Zoellick, 2012)

Since Rio+20, a variety of initiatives aimed at facilitating blue growth at global and national levels have been introduced. These have included the World Bank's Blue Growth Portfolio, the EU's Blue Growth Strategy, and Blue Growth programs in a wide range of countries in Africa and Australasia. The narrative of the blue economy as an economic imperative (good business) has oft been used as an argument in marine and coastal conflicts.

In the conflict over North Atlantic Mackerel, much of the discourse focused on the importance of fisheries to national economies. For Iceland, it was argued that, given the country was effectively bankrupted by the global financial crisis in 2008, returning to traditional industries and upping their catches was a logical step. 'Fishing is more important for Iceland than the car industry for Germany or the oil industry for Norway' (McDonald-Gibson, 2013). The rising value of mackerel had led to the mackerel fishing industry becoming big business with catches in Britain worth £135 million in 2009. Iceland believed the catch in 2010 would be worth 15 billion Icelandic krona (about £81 million), 'a big sum for a small country reeling from the financial crisis' (Economist, 2010).

The imperative for aquaculture and the dollars that it may bring into regional economies was clear in several case studies. In the British Columbia conflict between Marine Harvest and the First Nations indigenous groups, many believed that the growth of industry and the associated support from government is all about money into government coffers: 'The government does not have enough information to build a good regional strategy for fish farms. It's just all about the dollar in their decisions.' (Gerwing and McDaniels, 2006). In the same article, it was also pointed out that the First Nations were thinking about their own community economies 'First Nations are thinking strategically about how to enrich the community economic development opportunities that can bring stability to their people' (ibid.). In the case of Ecuador, aquaculture developments 'resulted from the government's need to meet the pressure for "developing" the country and reimbursing the external debt' (Veuthey and Gerber, 2012). In the Macquarie Harbour conflict in Tasmania, 'the success of the aquaculture industry it created from scratch 30 years ago is a perfect riposte to those mainlanders who paint the south isle as the "mendicant state"' (Denholm, 2016).

There are also indirect effects that changing uses of the marine environment can have on the economy. First Nations in Canada argued that should aquaculture developments increase, this would lead to a loss of wild salmon and thus also whales, grizzly bears and billions of dollars in local revenue from tourists: 'If we lose our wild salmon we lose our whales, we lose our grizzly bears, there goes that tourism industry' (Johnson, 2018). In the Cape Wind example, 'At 10 cents a month, do you risk all of this and all of what it means for our livelihood, for the tourism and the economy of Cape Cod?' (2005). In fact, the anti-development organisation – The Alliance – estimated that the project could result in a reduction of tourist spending of $57 million to $123 million.

Institutional failure

In the public policy realm, there are many examples of cases where important lessons remain unlearned, mistakes are buried, responsibility is deflected, and the causes of the original failure continue unchallenged. Often, the same – or at least similar – mistakes are repeated time and again. Institutional failure may be the reason that we have not got natural resource governance quite right yet. Institutional failure is often categorised into (i) law and policy failure, (ii) market failure, (iii) organisational failure (which may include system and structural failures), and (iv) informal institutional failures (e.g. breakdown of norms). We often see discourses around categories (i) and (iii) (those that relate directly to the roles of government agencies) arise in marine and coastal conflict. So, why do so many government efforts to manage ocean and coastal resources fail miserably?

Analysis of the conflict case studies revealed that there are three main issues: (i) non-existent planning or regulatory mechanisms, (ii) there has been a failure or breakdown of existing mechanisms, and (iii) there is a perception that evidence-based decision-making does not occur.

Many marine and coastal conflicts occur, in part, because either there are no planning or regulation mechanisms in place, or they are weak or poorly devised, or there is a perception that this is the case. In the Cape Wind conflict, opponents believed that there was no federal governing process or overview in place for the offshore wind farm,

> ... the Cape Wind project sets a dangerous precedent 'of a private developer being able to take over a public resource for free with no process in place to govern them.' He added: 'This deluge of proposals really points to the fact that we need some federal oversight over the use of the continental shelf. There is no fuel source that isn't regulated, except for wind power.'
>
> (Belluck, 2003)

In the case study of Crest Energy in New Zealand, 'There are no specific governance structures in place for MRE projects, and they are consented under a range of existing legislative instruments' (Kerr *et al.*, 2015: 112). Similarly, in salmon aquaculture conflicts in Australia, 'The fledgling industry and government is learning "on the job" what regulations are appropriate and, in some instances, these struggle to keep pace with the stellar growth' (Denholm, 2016). This may be the situation in all cases, given that the development of marine renewable energy and commercial aquaculture is still at an early stage. The industries themselves are driving growth and perhaps government agencies are not able to match the pace.

The failure or breakdown of existing mechanisms can also be a cause of marine and coastal conflict. It is evident that the breakdown of the existing management process for mackerel in the North Sea contributed towards the Mackerel Wars. 'The top-down system of TAC and quota-sharing has broken down ... At present a top-down multilateral TAC-based regulation has been replaced by the bottom-up unilateral and non-coordinated decision-making in the system of non-organized non-cooperative distribution' (Ørebech, 2013: 350). Part of the problem was that the mechanisms to incorporate newcomers into the agreements were either flawed or non-existent. In the case of aquaculture conflict in Ecuador, issues of overlapping mandates and sharing of responsibilities led to the failure of existing mechanisms,

> The intricate sharing of responsibilities in the management of mangroves and shrimp farming has virtually led to a legal vacuum allowing

> the illegal invasion and destruction of mangroves ... when complaints for illegal deforestation were lodged by local communities and NGOs, the various ministry organs evaded and shifted responsibilities.
>
> (Veuthey and Gerber, 2012)

Another aspect of the failure/breakdown of existing mechanisms that arose across many case studies is the perception that governments act outside of the existing process. This is particularly evident in those studies from Australia. In the aquaculture conflict around Okehampton Bay, Tasmania, it was stated that 'Marine farming happens outside the normal planning processes in this state' (Richards, 2017). As for the supertrawler conflict, it was suggested that decision-makers were not acting within the existing processes: 'AFMA should go through its process and this time act lawfully and do it properly' (Ockenden, 2012). This issue also cropped up in the South African conflict involving marine protected areas and fishers in Tsitsikamma National Park: 'the (environmental affairs department) is not responsible for fisheries, so it cannot allocate "fishing rights" under the Marine Living Resources Act, even in an MPA' (Kemp, 2010).

A final institutional failure which featured in several of the case studies concerns problems with what we actually know (or do not) and how this is used in decision-making. In the twenty-first century, decision-makers rely increasingly on science and expert knowledge and, as noted by Acheson (2006), 'an aura of certainty and infallibility has come to surround science and scientists' (p. 124). Yet the science around natural resource decision-making is not foolproof – it can be affected by poor research design, errors of measurement, and it is more often than not wrapped up in layers of uncertainty. Many groups are recognising these inadequacies and as a result do not support management measures, or they use them as a reason to campaign against changes they do not agree with. In the Lyme Bay case study, 'The fishermen say the Department for the Environment, Food and Rural Affairs has caved under pressure from the green lobby – and claim the science aimed at protecting the stunning rare reefs is flawed' (Vennells, 2008). In the case of the conflict between indigenous groups and Marine Harvest in British Columbia,

> A great deal of distrust of science has arisen as a result of outside 'experts' parachuting in, determining what is acceptable without consulting local First Nations and then leaving – while those who call these places home are left to deal with the consequences of decisions they had no part in making.
>
> (Page, 2007: 620)

Furthermore, the concept of the 'precautionary principle', when decisions to reduce or stop exploitation can be justified in situations with potential harm and extensive scientific knowledge on the matter is lacking, is frequently cited/applied in marine and coastal conflict. In the supertrawler case study 'Concerns from fishers about the potential for localised over-fishing and impacts on predators such as tuna remain unaddressed ... we welcome more scientific research into our fisheries' (Hubbard, 2014). In the case of Okehampton Bay opponents argued 'stop any expansion of the salmon industry until the very best minds do the research and determine the right way forward, a way that genuinely satisfies the most stringent environmental, social and economic criteria' (Wilkie, 2017). In some respects, this is an easy argument to make – after all we know less about the oceans than we do about outer space. As we move into a 'post-truth' era, in which facts are labelled subjective and any information that conflicts with our own opinion is treated with scepticism, we may see this argument even more frequently.

A changing marine and coastal environment

With an increasing population and the associated drivers of security of supply and the economic imperative, all engulfed within a framework of governance failure, we are seeing huge changes in our marine environment. Already, no marine area is entirely unaffected by human influence (Halpern *et al.*, 2008). The Millennium Ecosystem Assessment, a four-year research effort to measure the value of natural resources, concluded that 'The structure of the world's ecosystems changed more rapidly in the second half of the twentieth century than at any time in recorded human history, and virtually all of Earth's ecosystems have now been significantly transformed through human actions.' (Sarukhán *et al.*, 2005).

Across the conflict case studies, an emerging discourse of a changing marine environment, as caused by humans, was evident. In British Columbia,

> Nuu-cha-nulth have been listening to biologists and scientists for decades tell them that everything is fine, only to see their sea resources diminished to levels that will no longer support a people that are inextricably linked ... to these resources.
>
> (Page, 2007)

In Ecuador, local communities 'attribute a decline in their fishery to a loss of mangroves due to pond construction and the damaging effects of pond effluents' (Carter, 1999). In the Lyme Bay case study,

> What is happening here is the sickness of the planet being acted out before our very eyes ... We are making deserts of the seabed as surely as logging of ancient woodland and intensive agriculture have made deserts on land.
>
> (Western Morning News, 2006)

As a species we are becoming more environmentally aware. We are more likely to use natural building materials, to install solar panels and use energy-saving lightbulbs, and we are more likely to recycle. Environmental awareness is particularly evident in countries that are wealthier and more highly educated (Knight, 2016), although still more than a third of the world's adults have never heard of climate change (Lee *et al.*, 2015). In fact, climate change is a discourse that did not arise in the analyses, but which really should be included in any exploration of the drivers of marine and coastal conflict. It is likely to be of increasing importance as we consider the future.

Half a century ago, in January 1968, the Stanford Research Institute delivered a report entitled 'Sources, Abundance, and Fate of Gaseous Atmospheric Pollutants' to the American Petroleum Institute, a trade association for the fossil fuel industry. The report was one of the first to recognise the risks of burning fossil fuels and it warned that if this was left unabated, it could bring about temperature increases, melting of ice caps, and sea level rise (Robinson and Robbins, 1968). Those predictions proved to be correct. Going into the twenty-first century, the concentration of carbon dioxide (CO_2) in the atmosphere had risen from about 323ppm (at the time of the report) to 369ppm (Idso *et al.*, 2001). Global temperature had increased by nearly half a degree over pre-industrial averages. In October 2018, 50 years on, the Intergovernmental Panel on Climate Change (IPCC) released a special report on the impacts of global warming of 1.5 °C above pre-industrial levels (likely to be reached between 2030 and 2052 at the current rate). The report warned that:

> Global warming of 1.5 °C is projected to shift the ranges of many marine species, to higher latitudes as well as increase the amount of damage to many ecosystems. It is also expected to drive the loss of coastal resources, and reduce the productivity of fisheries and aquaculture (especially at low latitudes).
>
> (IPCC, 2018: 10)

The way climate change affects the ocean also furthers risks to fisheries and aquaculture via impacts on the physiology, survivorship, habitat, reproduction, disease incidence, and risk of invasive species. On top of this, the level of ocean acidification due to increasing CO_2 concentrations associated with global warming of 1.5 °C is projected to amplify the adverse effects of warming, further impacting the growth, development, calcification, survival, and thus abundance of a broad range of species from algae to fish. There is also the potential for climate-related risks to health, human security, and economic growth. Populations at disproportionately higher risk include disadvantaged and vulnerable populations, some indigenous peoples, and local communities dependent on coastal livelihoods – particularly those living in small island developing states. The IPCC report warned that increasing warming amplifies the exposure of small islands, low-lying coastal areas, and deltas to the risks associated with sea level rise including increased saltwater intrusion, flooding and damage to infrastructure.

Summary

This study has identified five drivers of marine and coastal conflict: population growth; the need for security of supply of water, food and energy; the imperative for blue growth; institutional failure; and a changing marine environment. Changes to the marine environment could drive conflict where security of supply will become a greater concern, compounded by an increasing population and more migration to escape flooding or even drought. The economic imperative will continue to be a driver, but will certainly interact with changing distributions of species and changing water temperatures and quality. The impacts of all the above are likely to be exacerbated by institutional failure if governing institutions are unable to keep up with the pace of environmental change. In an era of huge environmental change, the drivers of conflict described here really are wicked problems; that is, due to their complex interdependencies, they are difficult or impossible to solve because of incomplete, contradictory, and changing requirements amplified by the non-existence of models or tools to assess, analyse, and extrapolate potential solutions.

Combined, these factors are creating competition between users for access to marine natural resources. Healthy competition can be a good thing. It can provide motivation for us to work hard and achieve objectives. It can engender greater commitment to quality as we seek to use this as a lever to beat our competitors. Competition can also serve as a spark to creativity as we seek out all the possible avenues to get ahead of the crowd. However, competition also has many cons. It can lead to compromised values and resentment. Competition can also turn up the volume on inequality, turning competition into conflict.

References

Acheson, J. M. 2006. Institutional failure in resource management. *Annual Review of Anthropology*, 35, 117–34.

Aquaculture New Zealand. 2011. *Overview* [Online]. Available: www.aquaculture.org.nz/industry/overview/ [Accessed 11 April 2018].

Belluck, P. 2003. A wind power plan stirs debate in Massachusetts. *New York Times*, 2 March.

Bowen, M. G. and Power, F. C. 1993. The moral manager: Communicative ethics and the 'Exxon Valdez' disaster. *Business Ethics Quarterly*, 3, 97–115.

BP. 2017. *BP Energy Outlook* [Online]. Available: https://safety4sea.com/wp-content/uploads/2017/01/BP-Energy-Outlook-2017_01.pdf [Accessed 11 March 2019].

BP Energy Economics. 2018. *BP Statistical Review of World Energy*. 67th edn. Available: www.bp.com/content/dam/bp/business-sites/en/global/corporate/pdfs/energy-economics/statistical-review/bp-stats-review-2018-full-report.pdf [Accessed 11 March 2019].

Carter, E. 1999. *Narratives of Environmental Change: Mangroves, Shrimp Farming and Artisanal Fishing Communities in Ecuador*. Master's thesis, University of Wisconsin-Madison. Available: https://minds.wisconsin.edu/handle/1793/47106/ [Accessed 22 March 2019].

Cleland, J. 2013. World population growth: Past, present and future. *Environmental and Resource Economics*, 55, 543–54.

Country Garden Malaysia. 2015. *Forest City project recommences as Country Garden Pacificview prioritises environmental mitigation measures* [Online]. Available: https://cgpvforestcity.wordpress.com/category/press-release/ [Accessed 12 April 2018].

Crider, M. C. 2012. *The social and environmental effects of shrimp mariculture: Case studies of two coastal villages in Ecuador*, Florida Atlantic University.

Cruise Lines International Association 2018. Travel Agent Cruise Industry Outlook Winter 2018. CLIA Global.

Denholm, M. 2016. Dangerous dinners: Is salmon farming ruining Tasmania? *The Australian*, 22 December.

Desvaux, M. 2007. The sustainability of human populations. *Significance*, 4, 102–7.

Economist. 2010. Mackerel wars: Overfished and over there. *Economist,* 2 September.

Faasen, H. and Watts, S. 2007. Local community reaction to the 'no-take' policy on fishing in the Tsitsikamma National Park, South Africa. *Ecological Economics*, 64, 36–46.

FAO. 1996. *Rome declaration on world food security* [Online]. Available: www.fao.org/3/w3613e/w3613e00.htm [Accessed 13 April 2018].

FAO. 2009. How to Feed the World in 2050 [Online]. Available: www.fao.org/fileadmin/templates/wsfs/docs/expert_paper/How_to_Feed_the_World_in_2050.pdf [Accessed 11 March 2019].

FAO, IFAD, UNICEF, WFP, and WHO. 2017. *The state of food security and nutrition in the world 2017*. FAO.

Gerland, P., Raftery, A. E., Ševčíková, H., Li, N., Gu, D., Spoorenberg, T., Alkema, L., Fosdick, B. K., Chunn, J., and Lalic, N. 2014. World population stabilization unlikely this century. *Science*, 346, 234–7.

Gerwing, K. and McDaniels, T. 2006. Listening to the salmon people: Coastal First Nations' objectives regarding salmon aquaculture in British Columbia. *Society and Natural Resources*, 19, 259–73.

Global Wind Energy Council 2017. Global wind 2017 report: A snapshot of top wind markets [Online]. Available: www.tuulivoimayhdistys.fi/filebank/1191-GWEC_Global_Wind_Report_April_2018.pdf [Accessed 10 March 2019].

Halpern, B. S., Walbridge, S., Selkoe, K. A., Kappel, C. V., Micheli, F., D'Agrosa, C., Bruno, J. F., Casey, K. S., Ebert, C., and Fox, H. E. 2008. A global map of human impact on marine ecosystems. *Science*, 319, 948–52.

Higgins, A. 2011. In South China Sea, an energy battle. *Fiscal Times*, 18 September.

HLPE 2014. Sustainable fisheries and aquaculture for food security and nutrition. *A report by the High Level Panel of Experts on Food Security and Nutrition of the Committee on World Food Security*. FAO.

Hoegh-Guldberg, O. and Bruno, J. F. 2010. The impact of climate change on the world's marine ecosystems. *Science*, 328, 1523–8.

Hubbard, R. 2014. *Fishers and conservationists seek election commitment to permanently ban super trawlers* [Online]. Available: www.stopthetrawler.net/election_campaign_launch [Accessed 11 January 2018].

Idso, C. D., Idso, S. B., and Balling Jr, R. C. 2001. An intensive two-week study of an urban CO_2 dome in Phoenix, Arizona, USA. *Atmospheric Environment*, 35, 995–1000.

IPCC 2018. Global Warming of 1.5 °C. An IPCC Special Report on the impacts of global warming of 1.5 °C above pre-industrial levels and related global greenhouse gas emission pathways, in the context of strengthening the global response to the threat of climate change, sustainable development, and efforts to eradicate poverty [Online]. Available: www.ipcc.ch/2019/05/13/ipcc-2019-refinement/ [Accessed 10 March 2019].

Johnson, R. 2018. *First Nations face off against open-pen salmon farms in British Columbia* [Online]. Available: www.earthisland.org/journal/index.php/elist/eListRead/First_Nations_face_off_open-pen_fish_farms_british_columbia/ [Accessed 23 March 2018].

Kemp, Y. 2010. New move to open marine park to fishers sparks fears. *Cape Argus*, 11 October.

Kerr, S., Colton, J., Johnson, K., and Wright, G. 2015. Rights and ownership in sea country: Implications of marine renewable energy for indigenous and local communities. *Marine Policy*, 52, 108–15.

Knight, K. W. 2016. Public awareness and perception of climate change: a quantitative cross-national study. *Environmental Sociology*, 2, 101–13.

Kobayashi, M., Msangi, S., Batka, M., Vannuccini, S., Dey, M. M., and Anderson, J. L. 2015. Fish to 2030: The role and opportunity for aquaculture. *Aquaculture Economics & Management*, 19, 282–300.

Lee, T. M., Markowitz, E. M., Howe, P. D., Ko, C.-Y., and Leiserowitz, A. A. 2015. Predictors of public climate change awareness and risk perception around the world. *Nature Climate Change*, 5, 1014.

McDonald-Gibson, C. 2013. EU tackles Iceland over 'mackerel wars'. *Independent*, 28 July.

Miga, A. 2006. Lawmakers reach compromise on wind farm. *Associated Press*, 22 June.

Murtaugh, P. A. and Schlax, M. G. 2009. Reproduction and the carbon legacies of individuals. *Global Environmental Change*, 19, 14–20.

O'Connor, S., Ono, R., and Clarkson, C. 2011. Pelagic fishing at 42,000 years before the present and the maritime skills of modern humans. *Science*, 334, 1117–21.

Ockenden, W. 2012. *Ombudsman to review super fishing quota*, [Online]. Available: www.abc.net.au/worldtoday/content/2012/s3572323.htm [Accessed 11 March 2019].

Ørebech, P. 2013. The 'lost mackerel' of the North East Atlantic – the flawed system of trilateral and bilateral decision-making. *The International Journal of Marine and Coastal Law*, 28, 343–73.

Page, J. 2007. Salmon farming in First Nations' territories: A case of environmental injustice on Canada's West Coast. *Local Environment*, 12, 613–26.

Radio New Zealand. 2011. *Protest at tidal power project planned* [Online]. Available: www.radionz.co.nz/news/national/72793/protest-at-tidal-power-project-planned [Accessed 13 April 2018].

Ramsar 2017. *Fact Sheet 3. Wetlands: A global disappearing act* [Online]. Available: www.ramsar.org/document/ramsar-fact-sheet-3-wetlands-a-global-disappearing-act [Accessed 11 March 2019].

Richards, B. 2017. Nod to fish farm jobs. *Hobart Mercury*, 10 July.

Robinson, E. and Robbins, R. 1968. *Sources, abundance, and fate of gaseous atmospheric pollutants. Final report and supplement.* Stanford Research Institute and American Petroleum Institute.

Sarukhán, J., Whyte, A., Hassan, R., Scholes, R., Ash, N., Carpenter, S., Pingali, P., Bennett, E., Zurek, M., and Chopra, K. 2005. Millennium Ecosystem Assessment: Ecosystems and human well-being [Online]. Available: www.millennium assessment.org/documents/document.788.aspx.pdf [Accessed 11 March 2019].

Sperry, T. 2012. Wind farm gets US approval despite controversy. *CNN Wire*, 17 August.

Statista. 2019a. *Container Shipping – Statistics & facts* [Online]. Available: www.statista.com/topics/1367/container-shipping/ [Accessed 11 April 2018].

Statista 2019b. Global fish production from 2002 to 2017 (in million metric tons) [Online]. Available: www.statista.com/statistics/264577/total-world-fish-production-since-2002/. [Accessed 11 April 2018].

Subasinghe, R. 2017. *World aquaculture 2015: A brief overview.* FAO Fisheries and Aquaculture Report.

Swyngedouw, E. 2013. UN water report 2012: Depoliticizing water. *Development and Change*, 44, 823–35.

U.S. Energy Information Administration. 2016. *Offshore production nearly 30% of global crude oil output in 2015* [Online]. Available: www.eia.gov/todayinenergy/detail.php?id=28492 [Accessed 11 April 2018].

UN Oceans 2008. *United Nations Atlas of the Oceans.* Available: www.oceansatlas.org/ [Accessed 11 March 2019].

United Nations 2017. The Ocean Conference Factsheets. Available: www.un.org/sustainabledevelopment/wp-content/uploads/2017/05/Ocean-fact-sheet-package.pdf [Accessed 11 March 2019].

Vennells, L. 2008. Fishing rights snatched away. *Western Morning News*, 20 June.

Veuthey, S. and Gerber, J.-F. 2012. Accumulation by dispossession in coastal Ecuador: Shrimp farming, local resistance and the gender structure of mobilizations. *Global Environmental Change*, 22, 611–22.

Western Morning News. 2006. New call for dredging ban. *Western Morning News*, 30 June.

Wilkie, A. 2017. Union's cynical campaign undermines our vital salmon industry. *Hobart Mercury*, 7 February.

WWF. 2016. *Living Planet Report 2016: Risk and Resilience in a New Era.* Available: www.worldwildlife.org/pages/living-planet-report-2016 [Accessed 11 March 2019].

Zoellick, R. 2012. A new SOS: Save our seas. *Keynote speech, The Economist's World Ocean Summit, Singapore,* 24 February.

Chapter 6

The rules, rights, and effects of marine resource use

The previous chapter revealed that there are several factors driving increasing use of the marine and coastal environment and that this leads to competition over these resources. The realisation of limitations to such natural resources stirs up the question of how best to manage them. To answer this question, we need to understand the rules, rights, and effects of marine resource use.

Marine governance inherently takes place in a highly complex social-ecological environment. It is a politically and culturally driven process, shaped by human livelihoods and perceptions, moulded by various rights and rules, and with myriad effects. As noted by Beyerl *et al.* (2016), governance is influenced by external factors (e.g. environmental change, market supply and demand) as well as internal community-specific conditions (e.g. inherited ownership structures, hierarchies, religious influences, or societal obligations). An individual's perception of these factors significantly shapes the dynamic of the entire governance process, including decisions on management measures and their execution. An individual's perception of these factors may also determine their involvement in marine and coastal conflict. In this chapter, I set out four sets of perceptions, identified during the conflict case study analyses. These perceptions directly or indirectly relate to the governance process and can be considered factors which exacerbate the potential for marine and coastal conflict.

Don't take or alter what is ours (or no-one's)

We perceive that marine and coastal resources belong to no one person, but rather to all of us (be that for our own society within our nation state territories, and globally within the high seas) or, to none of us. This is what we call a 'common pool' resource, as explained in Chapter 2. There are two important attributes to a common pool resource (Ostrom *et al.*, 1994): that exclusion of users is difficult and that by using a resource each user can detract from the welfare of others. The statement 'don't take or alter what is ours' refers to an often-verbalised belief that we have been excluded from a resource, or that our welfare has been affected by use of a resource by others. This belief can motivate us to act, oftentimes leading to conflict.

Exclusion

Excluding or limiting potential beneficiaries from using a commons such as the oceans and coasts is difficult, although not impossible. Fishing is one very clear instance of this. It would be impossible to fence off one fishery, never mind a whole ocean, and therefore there are many cases of illegal fishing. Any 'fencing off' of a resource must be backed by a set of property rights that are feasible to defend, and as we saw in Chapter 2, we often see either quasi-rights or a complete lack of property rights around marine and coastal resources. Issues around 'ownership' and 'rights' arose frequently across the case studies, at nation state and local community scales. Common pool resources may be owned by national, regional, or local governments as public goods, by communal groups as common property resources, or by private individuals or corporations as private goods. When they are owned by no-one (as is the case with the high seas), they are referred to as open access resources. This 'ownership' tends to be broken down into separable rights including access (right to enter), withdrawal (right to extract), management (right to regulate use), exclusion (right to deny access), and alienation (right to sell, lease, or transfer) (Ostrom and Schlager, 1996). Some individuals or groups hold all of these rights while others hold only some of them. This leads to confusion regarding ownership and rights.

The ways in which we understand claims to marine resources have changed over time. In 1609, Grotius's concept of Mare Liberum (Grotius, 1916) posited that the sea should be free to all, that nobody had any right to deny any other access. Grotius' argument was influential but was countered by arguments from de Freitas (Portugal) and Selden (England) (Knight, 1925, Selden, 1751). For some time after this, rights to the sea were limited to three nautical miles (nm) from the coastline – the maximum distance a cannonball could be fired (Zacharias, 2014). Waters beyond national boundaries were considered open to all nations, but the property of none. The issue of claims to marine waters was raised again with the United Nations in 1967. This led to the United Nations Convention on the Law of the Sea (UNCLOS), which came into force in 1994, and defined the rights and responsibilities of nations with respect to their use of the world's oceans. UNCLOS resolved several issues related to ocean usage and sovereignty: establishing freedom-of-navigation rights; setting territorial sea boundaries to 12 nm offshore (22 km), in which the coastal state is free to set laws, regulate use, and use any resource; setting exclusive economic zones (EEZs) up to 200 nm offshore (370 km) in which the coastal nation has sole exploitation rights over all natural resources; and establishing rules for extending continental shelf rights up to 350 miles offshore (650 km).

Regardless of the clarifications made by UNCLOS, disputes between nations regarding maritime resources are very much evident today. For example, certain areas of the Arctic sea region are in dispute; Canada, Denmark, Norway, Russia, and the United States all consider parts of this area as their national waters. There is an ongoing dispute involving a small slice of the Beaufort Sea between the Canadian territory of the Yukon and the United States state of Alaska. There are also disputes regarding what passages constitute international seaways and rights to passage along them, including the North-west Passage, which Canada considers internal waters and the United States and most other maritime nations consider an international transit passage. To add some comedy to such disputes, in 2007 Russia symbolically staked its claim to oil and gas reserves in the Arctic Ocean by planting a titanium flag on the seabed under the North Pole. The issue of rights is at the heart of arguments around the Spratly Islands conflict where 'all disputes are centered on sovereignty and rights over the Nansha Islands and their surrounding waters' (Manila Bulletin, 2016).

Exclusion-related conflicts also occur between users. In fisheries, two main forms of use rights lead to exclusion:

- access rights, which permit the holder to take part in a fishery (limited entry) or to fish in a location (territorial use rights);
- withdrawal rights, which typically involve quantitative (numerical) limits on resource usage, either through input (effort) rights or output (harvest) rights.

Access rights refer to the capability to enter a fishery and to participate in it. These are often subject to several constraints such as how, when, and where participation is allowed. The two best-known forms of access rights are territorial use rights in fishing (TURFs) and limited entry licences. TURFs allocate secure, exclusive privileges to fish areas to groups (often in the form of co-operatives), or in some rare cases, to individuals. The territory governed by a TURF can relate to the surface, the bottom, or to the entire water column within a specific area. They have often emerged from centuries-old traditions and in some countries, such as Vanuatu, Papua New Guinea, and the Solomon Islands, these traditional rights are recognised by national law. While mostly associated with developing regions, TURFs can also be found in Japan and in the lobster fisheries of North America. Alternatively, limited entry licensing does exactly that, it limits the number of licences to fish. An existing licence may be transferred to another owner, but no new licences may be issued. Interestingly, many marine and coastal conflicts involve fishers' claims regarding a loss of spatial access to a resource, and yet this tends to occur in locations where no spatial access rights exist, rather the fishery is managed through withdrawal rights management tools.

Total allowable catch (TAC) is a withdrawal right that limits the total output from a fishery by setting a maximum weight or number of fishes that can be harvested over a period. It is based on indicators of biological production and is set by the regulator. Individual transferable quotas (ITQs) are a share of the TAC, allocated to individuals, which can be leased or traded. Total allowable effort (TAE), used less widely than TACs and ITQs, is a second form of withdrawal right. The TAE is typically denominated in nominal units of effort such as days at sea or sets of gear. Both types are essentially a de facto privatisation of a public resource, which provide stronger exclusive use of the right by individual firms, vessels, or groups. The Mackerel Wars conflict revolved around the size and relative allocations for the Atlantic Mackerel TAC. With the stocks moving northward into Iceland's EEZ and Iceland unilaterally increasing their TAC, if the stocks were to be kept sustainable, the other European countries TACs would have to be reduced. If this happened, then the other European countries would have lost their resource use rights.

Exclusion rights also exist regarding access to, and use of, the foreshore and seabed. Increasing industrial use of marine resources, such as energy extraction or aquaculture, has led to a struggle for further enclosure (over and above EEZs), analogous to that experienced with the privatisation of agricultural lands in England in the sixteenth century and the rest of Europe in the nineteenth century. In feudal times, land was owned by the nobility. Peasants used land that wasn't theirs to grow food to survive. In return for using the lord's land, they owed duties to the lord. However, during industrialisation, the commons were extinguished and transformed into private property owned by individuals. Wealthy nobles and lords decided that they should have the sole right to own certain pieces of land and should be able to use it whenever and however they wanted. To enclose land was to put a hedge or fence around a portion of this open land and thus prevent the exercise of common grazing and other rights over it. Of course, this is not possible in the marine environment, but states can provide the legal right to use a geographically defined area of the sea floor. Property rights for such uses tend to be well-defined spatially and temporally, often include exclusion rights, and once these uses are established, their rights tend to supersede those of other interests. This geographical exclusion, of the foreshore and seabed, can lead to conflict.

Marine Protected Areas (MPAs) are a more recent form of enclosure. MPAs are an area of sea especially dedicated to the protection and maintenance of biodiversity, and of natural and associated cultural resources, which is managed through legal or other effective means. MPAs come in a variety of forms. The International Union for Conservation of Nature offers seven categories of protected area: strict nature reserve (maximum protection), wilderness area, national park (light use allowed), natural monument or feature, habitat/species management area, protected seascape, and sustainable use of natural resources. In 2010, parties to the Convention on Biological Diversity (CBD) agreed to a target to protect over 10 per cent of the world's marine areas by 2020, now part of Sustainable Development Goal 14. Dependent on the type of MPA, there may be restrictions to certain activities. In some MPAs there may be limits to the catching of fish, collection of shells, or other activities where something may be extracted from the area. As with enclosure for industrial purposes, MPAs may also lead to the loss or restriction of resource use rights and this can lead to conflict.

Public access to marine resources is the point at which the issue of ownership and rights becomes more complicated. The fundamental right of public access to fisheries in tidal waters, for example, is well established. It existed in English law prior to the Magna Carta (which established for the first time the principle that everybody was subject to the law), where it went hand-in-hand with the doctrine of 'capture', which held that there was no ownership of swimming fish until they were captured. The same is true for the public right to navigation. But with discoveries of important resources and a sharp rise in all ocean uses generally, such accepted principles are rapidly beginning to erode. Customary law, for example, can no longer move fast enough to provide solutions to new problems. Of course, the nation state 'owns' the marine environment, as defined by UNCLOS, on behalf of its constituents. However, despite a perception that the public have rights to the oceans and coasts, and the resources that they provide – for example, in the Cape Wind conflict, 'There is almost a sense of ownership of the near ocean' (Kempton *et al.*, 2005) – there is no existing legislation which enforces this. Instead, it is often a strong cultural and historical association with the sea that leads to a perception or assumption of the existence of rights.

The belief of public rights to marine resources has contributed to increasing debate, and legislation, particularly regarding indigenous rights to sea country. In Australia, indigenous people's inherited rights to land were recognised by Australian law in 1992 with the Mabo High Court decision which became statutory law through the Native Title Act 1993. In New Zealand, the 1840 Treaty of Waitangi ensured that native title survived colonisation. However, Māori use of the foreshore and seabed has not always been recognised by the Crown. In 1997 an application was made to the Māori Land Court for determination of the foreshore and seabed in Marlborough Sounds as Māori customary land. The case was overturned, and the government passed the Foreshore and Seabed Act 2004 ensuring public rights of access and for activities taking place since 1840. In 2011, however, the act was replaced with the Marine and Coastal Area (Takutai Moana) Act 2011. This provided a 'no ownership' regime for the foreshore and seabed and meant that Māori could apply for recognition of customary rights. The Royal Proclamation of 1763, and the many treaties which it subsequently underpinned, have ensured that Canadian 'Aboriginal title' is recognised. There is also a comprehensive claims system in parts of the country where native title has not previously been dealt with by treaty or other means. Indeed, some First Nations groups claim that 'the provincial and federal governments do not have the right to grant [e.g. aquaculture] companies' tenures in their lands and waters without their consent' (Gies, 2017).

Subtraction

The second issue, subtraction, can be understood as the reduced quantity or quality of a resource. If one fisherman lands a tonne of fish, then those fish are not available for another fisherman to catch. When exposed to an oil spill, the oil pollution may kill fish in contained or near shore areas, meaning they are no longer available to catch. The installation of an offshore wind turbine array may mean that a person can no longer have an aesthetic appreciation of a seascape, or no longer sail a recreational yacht through that area. Resources can be degraded (decline to a lower quality) or depleted (decline to a lower quantity faster than replenished) in two key ways: through the direct consumption of a resource or through activities that cause indirect harm to the resource (often referred to as 'negative externalities'). Between them, degradation and depletion produce a decrease in the availability of the resource supply, meaning that each person's slice of the resource pie is smaller. This means that we see more conflict over who gets what little pie there is.

With over 70 per cent of the world's fish species either fully exploited or depleted as of 2005 (FAO, 2005), conflict often arises in response to subtraction in the form of declining fish stocks and shrinking catches. Overexploitation does not only cause these problems for target populations but can also result in indirect effects such as the trophic cascading effects identified when top-level predators are removed from the ecosystem. In the Bay of Bengal, fisheries have been under pressure for decades and are now severely depleted. In a bay that was once feared by sailors for its man-eating sharks, apex predators have been badly hit. This was largely precipitated by western aid, leading to the 'pink gold rush' for prawns in the 1960s, as well as the modernisation and expansion of the small-scale fisheries sector following the tsunami in 2004. As catch rates began to decline in the late 1990s, fishers were forced to move farther and farther from their home waters, leading to tensions between fishers from the various countries encircling the bay. In 2015 Sri Lankan authorities claimed to have spotted 40,544 Indian trawlers in Sri Lanka's territorial waters. Seventy trawlers were seized, and 450 fishermen were arrested. At least 100 deaths have been reported (Ghosh and Savio Lobo, 2017).

Conflicts may also erupt between wildlife protectionists and fishing industries, or between commercial and recreational fisheries due to subtraction. For example, fear of the potential for overfishing may lead to conflict. In the Supertrawler conflict, conservationists saw boats such as the supertrawler as a key cause of overexploitation. 'Super trawlers like the Margiris are one of the greatest drivers of overfishing – there are too many oversized boats chasing too few fish' (Hubbard, 2012). They argued that the size of the vessel would lead to a localised depletion of fish stocks as well as by-catch of marine mammals such as seals and dolphins. An argument was made that the ship was kicked out of West Africa for almost exterminating small pelagic fish stocks, and that 'its Dutch-owned sister ships were among the main culprits responsible for fishing down the once bountiful stocks of Peruvian Jack mackerel to commercially extinct levels in only about four years' (Wainwright, 2015).

Subtraction in the form of pollution is a second major theme in marine and coastal conflict. Pollution is the introduction of contaminants into the natural environment, which can cause adverse change. Pollution can take many forms. In the ocean and coastal environment, this includes water pollution caused by the discharge of wastewater from commercial and industrial waste, discharges of untreated sewerage or chemical contaminants from treated sewerage, and surface run-off flowing into surface waters (such as urban and agricultural run-off) or marine debris, which may have been deliberately or accidentally released into the environment, often including non-biodegradable plastics. In the case of the Exxon Valdez Oil spill, it was believed that the spill pollution had spoiled the beauty of Prince William Sound in Alaska, an area which capitalised on its beauty through tourism and arts. The effects of visual pollution were a large issue in the Cape Wind conflict. Many opponents of the Cape Wind project argued that the turbines would be ugly, unsightly and a blight on the coastal viewshed. There were concerns about 'the project's "destroying the scenic value of the sound," not wanting their view "spoiled" and concern about the impact on their property values' (Kempton *et al.*, 2005: 137). However, others held the belief that they were like artwork, or were beautiful because they were symbolic of doing the right thing. Finally, aquaculture has been perceived as releasing pollution daily into the ocean which has been viewed as a 'direct assault on the survival of several wild fish species' (Morton, 2017). Arising time and again is the concept of 'pristine nature'. This is the idea that the ocean should be kept free from human intrusion and that ocean and coastal areas are clean, clear, untouched, and unspoiled. In the Brent Spar conflict: 'that of the sanctity of the deep ocean. One should not dump in it as it supposedly has not been dumped in before. It should remain pristine and untouched' (Löfstedt and Renn, 1997). Similarly, after the Exxon Valdez oil spill, one resident noted that ' "We had a pristine environment," he said. "What we've got now, well, it's a loss of innocence." ' (Egan, 1989).

Habitat loss, caused by destruction of ocean and coastal areas for development, is also a form of degradation and thus subtraction. Coastal areas, with their closeness to human population centres, have suffered disproportionately; these critical areas, which include estuaries, swamps, marshes, and wetlands, often serve as breeding grounds or nurseries for nearly all marine species. We dredge and fill in wetlands to accommodate urban, industrial, and agricultural development. Indeed, it is estimated that every day between 1960 and 1995, 1 km of coastline was developed, causing permanent losses of valuable habitats, such as coastal wetlands, seagrass meadows, and rocky shores (Airoldi and Beck, 2007). Furthermore, deforestation far from shore has created erosion, sending silt into shallow waters blocking the sunlight that coral reefs need to thrive. Habitat loss has been generally associated with drastic declines in overall abundance and diversity of marine organisms. It has also been at the heart of several conflicts. In the dyke construction project conflict arose between demands for coastal protection including work on dykes and excavating clay and sand for use as building material on one hand and demands for preserving and maintaining the saltmarshes on the other. In the Muisne Canton in Ecuador, shrimp mariculture has been strongly opposed by locals and environmental associations, who engaged in defending the mangrove ecosystems in which the shrimp cultivation ponds were located.

Don't treat us unfairly

Since the Love Canal tragedy (Beck, 1979), the environmental justice movement has grown, with significant advances in legislation, legal cases, scholarship, and activism. Those who are concerned with environmental justice view the environment as a form of goods and bads that a society must divide among its members; but believe that the poor suffer a disproportionate amount of the bads while the rich enjoy a disproportionate amount of the goods. Activists and researchers have identified and challenged stark spatial patterns of environmental injustice where chemical factories, toxic waste sites, and other environmental hazards have been deliberately sited among poor, marginalised, and vulnerable populations (Melosi and Stine, 2006, Williams, 1999, Cifuentes and Frumkin, 2007). Justice provides local communities and environmental activists with an important terminology in their resistance against dispossession from customary land, opposition to polluting industries and struggles for a fair distribution of natural resource revenues. Claims of environmental injustice have increasingly become part of environmental conflicts, both explicitly through the work of campaign groups and implicitly through the arguments deployed about the rights and wrongs of a given situation. Such claims centre on different notions of justice, related to the available quantities of goods or bads, the process by which goods or bads are to be distributed, and the resulting allocation of the goods or bads to various factions of society.

Distributional justice

As the twenty-first century global economy increasingly hits natural resource limits, fundamental questions about 'fair shares' are beginning to emerge. Distributive justice refers to whether we obtain these fair shares. It is a complex concept with many different understandings, all of which vary in what is considered relevant (income, opportunities, jobs, welfare, etc.); in the nature of the recipients (individual persons, groups of persons, etc.); and on what basis the distribution should be made (equality, according to individual characteristics, according to free transactions, etc.).

Without a shared societal understanding of distributional justice, or a clearly articulated goal as to how to use our marine spaces (as is often the case with the marine and coastal environment), fairness judgements are left to the perceptions of individuals. Deciding what is a fair share of a resource tends to be done by means of social comparison. We look at other people's outcomes as a reference point for judgement of our own outcomes. However, what is perceived as a fair distribution will not be the same for different types of goods (or harms). As an example, at work we commonly expect to be rewarded in proportion to our skills and effort, but when we are sick we expect to get the correct treatment regardless of skills and effort.

A fairness judgement is something that we feel, rather than something that is rationally explained, particularly *un*fairness. Such judgements are based on our morals – deeply held beliefs that, if violated, will lead to a perception of unfairness, no matter how fair the process or decision was. What's more, values and beliefs are not static, they change over time. A key problem with the concept of distributive justice is that we tend to make self-serving judgements about what we deserve. For example, we overestimate our own performance and consequently what we are entitled to. When the situation is ambiguous, this tendency is intensified, and we are more likely to over-allocate resources to ourselves. Natural resources are heterogeneous in nature and so different people and different societies will value them differently. This means that the way we interact with the ocean and coastal environment results in complex dilemmas regarding the distribution of marine and coastal resources.

Regardless of any sense of injustice felt at the level of the individual person, for conflict to occur, the sense of injustice is (most often) experienced at a group level. As noted by Lind *et al.* (1998), the most powerful social injustice judgements tend to involve a shared and mutual reinforcement of injustice experiences. According to social identity theory, we tend to behave in ways that favour our own group and discriminate against other groups (Gao *et al.*, 2008). Therefore, if we feel that the group we belong to has been treated unfairly, this may exacerbate the potential for conflict.

Fairness and equity arose as themes across the marine and coastal conflicts studied here. For example, the basis of the North Atlantic Mackerel wars conflict was underpinned by the idea of negotiating a fair share of the bounty. In the conflict on aquaculture in British Columbia, Canada, there was the perception that 'There is only a handful of people that actually benefit from this cancer along our coast' (Johnson, 2018). In Tsitsikamma National Park, where much of the marine space was closed to fishing, the local communities believed that the benefits of the park only flowed to a few and communities sought a more equitable sharing, with the park managers, of the benefits. In the case of shrimp farming in Ecuador, shrimp farmers used public infrastructure such as roads and docks yet were accused of not contributing to the upkeep of such infrastructure. It is widely accepted that 'free riding' by actively taking collectively produced goods without paying can be wrong, because it is unfair.

Environmental injustices refer not only to distributions of environmental hazards and amenities within bounded localities, but also to the ways in which place-specific policies and practices can have consequences that cross national boundaries, affect multiple scales, and extend across global networks. Even if intra-local practices or distributions can be conceived as 'just' according to one or another set of criteria, they can result in unjust extra-local outcomes. In the Cape Wind case, the perspective is that people of Cape Cod would receive the brunt of the impact of the project, but not the benefits. 'Cape Wind would be industrialising one of the world's last great natural resources and profiteering at the expense of cape residents' (Noble, 2002), the argument being that the energy and the corporate profits would be shipped elsewhere. This is a common argument in marine and coastal conflicts, particularly those which relate to industrial use of the marine environment. Oftentimes, developments relating to e.g. aquaculture or to energy extraction are established in rural/regional locations, mostly for reasons of the high quality of the resource in those locations (e.g. cleaner or faster flowing waters), but also sometimes for reasons of public acceptability. However, for those that live in such locations, they often perceive that they have to deal with the impacts (which may be biological, chemical, visual, or noise pollution, among other 'bads') while the benefits flow outside of the locality. These flow-out benefits relate to where the value of production lies and may be in the form of supply chains and customers located in more urban locations or inter/multi-national ownership and shareholders.

Procedural justice

Given that the environment cannot sensibly be experienced equally or uniformly, it can become positively perverse to seek its 'even' distribution (whatever that might mean). In this sense, an unequal distribution of environmental goods or bads by itself may not necessarily be unjust, rather it is the 'fairness' of the processes through which the distribution has occurred which is important. When a process is viewed as 'fair', people are more likely to accept 'negative' outcomes.

People judge the fairness of a process in a few ways. These include: (i) consistency across persons and across time, (ii) suppression of personal self-interest, (iii) use of accurate information, (iv) modifiability of decisions, (v) representativeness of all recipient concerns, (vi) adherence to prevailing ethical and moral standards, (vii) neutrality, (viii) trust in the benevolent intentions of decision-makers, and (ix) status recognition (Lind and Tyler, 1988). The provision of relevant information, which promotes feelings of efficacy, can promote the perception that procedures are fair. Providing individuals with the opportunity to voice their concerns regarding natural resources leads them to believe that they have been treated more fairly, whether or not their input influences the decision. If their views are being considered, judgements of fairness increase even further (Lind *et al.*, 1990, Smith and McDonough, 2001). Many of these principles relate directly to the public participation process in natural resource decision-making. Although stakeholder and public views are often solicited, this is most often in the form of comment submission or through attendance at local meetings, and as such participation is limited. Not only have the content and substance of stakeholder and public input been constrained and removed from integration into planning processes, but the way input has been obtained has also operated to reduce understanding and has contributed to dissatisfaction and a sense of unfairness (Lawrence *et al.*, 1997). With increased used of the internet, there is the potential to access web-based environmental information, and the deliberative possibilities of virtual participation. However, as pointed out by Walker (2009), even these methods are, in practice, spatially and socially differentiated because the need to access these resources and the time–space constraints of everyday life limit the ability to be present in virtual participatory spaces. Several other challenges also limit progress towards achieving procedural justice (George and Reed, 2017). Incorporating a diversity of beliefs and values is difficult and as such the 'usual suspects' are those most often engaged with. Asymmetries in social power means that elite interests tend to be favoured. Also, because questions of sustainability have arisen out of the environmental movement, this has led to a focus on environmental priorities.

Procedural justice problems arise regularly in marine and coastal conflicts. As an example, the Cape Wind conflict involved protracted regulatory and judicial review which kept the project in limbo for some time. In the Macquarie Harbour conflict, decisions made by the Environment Protection Agency (EPA) regarding fish stocking densities led to a perception of an unfair distribution of the resource between the companies: 'the

14,000 cap was not being applied evenly to all three producers, giving Tassal a commercial advantage and costing Huon Aquaculture millions of dollars' (Whiting and Salmon, 2017). In a submission to the EPA in 2017, Huon argued that the EPA Draft Determinations (for levels of salmonid production in Macquarie Harbour) 'provide the other operators in MH with a commercial advantage over Huon and in effect, punishes Huon for its lower risk tolerance and prudent management of MH while rewarding the higher risk strategies of other operators' (Huon Aquaculture Group Pty Ltd, 2017) and that there was an 'inequity in the allocation method'. This was based on changes to the allocation process and the fact that Huon had voluntarily reduced their production to address the environmental impacts in the Harbour.

David and Goliath: power asymmetries in marine resource management

The portrait of a struggle between two sides arose frequently across the case studies analysed – in some cases between rich and poor, and in others between big business or big government and the public. In the Cape Wind case study, the conflict was described as 'very symbolic of the struggle between the rich and the public, the people' (Kempton *et al.*, 2005). The concept of David and Goliath was invoked on a number of occasions. Goliath is described in the biblical Book of Samuel as a tall Philistine warrior who was defeated by young David in single combat. More recently, however, the phrase has become popular to describe an underdog-type situation, one where a smaller and weaker opponent takes on a bigger and stronger adversary. In the Lyme Bay case study, 'Fishermen have resigned themselves to a "David and Goliath" struggle to protect their livelihoods and communities' (Herald Express, 2006). The Brent Spar case study also provides a great example of this in terms of the victory of Greenpeace over the multinational giant Shell (despite Greenpeace being no small organisation) being heralded by the media: 'Greenpeace, David, with its brave activists who occupied the platform, "slew" the big villain, Shell, Goliath' (Löfstedt and Renn, 1997: 133). In this struggle, big business (through huge power and influence) and big government are perceived as making decisions about how public goods are used, in ways which tend towards inequity.

In marine and coastal conflicts, we often see changing distributions of power over the period of the conflict. This is particularly clear in the Spratly Islands dispute where we have observed increasing naval power to the advantage of China, as well as an extension of capabilities on the reefs and islands that the country currently controls. We also see this in conflicts involving Indigenous groups who have started in a position of little power, but on invoking native rights, which are often enshrined in legislation, the power balance has shifted. As noted by Cascao and Zeitoun (2010), not all forms of power are equal. For example, material (economic, military, technological, or international support) and bargaining (capability of actors to control the rules of the game and set agendas) power may count for more than geographic position or ideational (capacity to impose and legitimise certain ideas or narratives) power.

Don't threaten our wellbeing

Human health and wellbeing are tied to the vitality of marine and coastal systems. We engage with these extraordinary environments to enhance our health and our wellbeing, and when these environments are affected in some way, accordingly it affects us, as individuals and as communities. Individual wellbeing is a combination of a person's physical, mental, emotional, and social health factors. Community wellbeing is the combination of those social, economic, environmental, cultural, and political factors essential for them to flourish and fulfil their potential.

Connections have already been made between human activity in the oceans and coasts, and health and wellbeing effects, both positive and negative. For example, climate change and extreme weather events have been linked to injury and mental health; anthropogenic chemicals have been found to link to cancer; seafood production has been linked to nutrition; and coastal communities have been linked to cultural history, employment, and commerce (Fleming *et al.*, 2015).

Threats to the individual

The loss of subsistence is a key concern for many when it comes to ocean and coastal resources. Many, including those in Tsitsikamma National Park, in Alaska and in Canada, raised concerns regarding the continued availability of the resources they depend on for survival. After the Alaskan oil spill, 'many of the natives of the region are afraid to eat the food they hunt at traditional hunting grounds such as Sleepy Bay. They claim a subsistence style of life that's thousands of years old has been irreparably damaged' (Sullivan, 1991). There is, however, the question of how many communities still rely 100 per cent on wild foods for survival. What may be a larger issue here is the relationship with culture and identity, and a history of subsistence.

In the case of a loss of, or restricted access to natural resources, many marine users feel that they are likely to see an impact upon their livelihoods. This concern tends to be raised by the fishing sector. At the time of the Exxon Valdez oil spill, 'Fishermen who ply Alaska's Prince William Sound for their livelihood are in deep trouble, and they know it' (Editorial, 1993). When Lyme Bay was closed to fishing, industry leaders predicted that the decision would 'act as the "final straw", driving some fishermen out of business' (Vennells, 2008). In Cape Cod, with the wind energy project, fishers argued that 'Horseshoe Shoal has been a prime fishing area for hundreds of years. This plan would destroy our livelihood' (Gallagher, 2002). Moving away from the fishing focus, in Ecuador, for the coastal populations who use mangroves for fuel, charcoal, tannins, building materials, and medicines, 'Mangrove destruction may lead to a decrease in job opportunities' (Crider, 2012). Livelihoods have also become the focus of arguments for changing access to resources. For example, in the New Zealand case study, it was argued that 'Northland's resources are underutilized, and such development will create jobs and therefore a sustainable economy for Northland to the benefit of the nation' (Simon, 2012). Similarly, in Cape Cod, 'Cape Wind is expected to add $1.5 billion to $2 billion to the United States economy and create 391 full-time jobs for operation and maintenance activities' (Electric Utility Week, 2004). However, a pertinent point is raised in the Okehampton Bay aquaculture conflict: 'they're looking at the few jobs that will be created here, but we forget about the jobs that will be lost in tourism' (Fantin, 2017). This brings to the fore the issue of precedence. Jobs and livelihoods are a focal point in almost all conflict case studies and are often argued about by both sides.

Health and safety are also issues that arise frequently in marine and coastal conflicts. Several aquaculture conflict discourses focused on the 'threat to health' that aquaculture can create:

> Farmed salmon are treated with antibiotics to combat disease, with chemicals to combat parasites and with artificial dyes to give them the characteristic pink flesh colour. Environmentalists and others worry that these treatments remain within the flesh of farmed salmon, and that they thus pose a food safety risk.
>
> (Page, 2007: 615)

They also focused upon the loss of recreational use that marine developments can pose and the effect on wellbeing that this can have. Regarding safety, boats need room to manoeuvre and new installations in the marine environment are likely to be a hazard, particularly at night or in poor weather. In the conflicts involving protected areas, safety issues arose from the requirement to travel further out to sea. In Lyme Bay, 'fishermen operated small vessels which were unsuited to deep waters, and lives could be endangered if they ventured too far to make a living' (Vennells, 2008). In the German dyke conflict, safety was more important to many participants in the conflict than the change to wetlands that the structure would cause.

We are aware of the clear links between natural environments and mental health. Studies have shown that spending time in or being active in natural environments is associated with positive outcomes for fatigue, anger, and sadness, as well as physiological stress. Larger areas of natural environment around where we live has also been associated with reduced risk of stress, psychological distress, and depression (Thompson *et al.*, 2012, Thompson Coon *et al.*, 2011). Therefore, an expanding human dominated landscape (or in this case seascape) caused by increasing use and industrialisation of the marine environment, thus leading to a loss of/reduction in natural space, may affect our mental health. A clear example of some of the effects that changes to environmental resources can have on the individual, thus leading to conflict, can be seen in the case of the Exxon Valdez oil spill in Alaska (Palinkas *et al.*, 1993). Losing our livelihood, or mental health problems may remove us from a group with which we identified. Our family may have been fishers, and our friends may have been fishers, but suddenly we are no longer able to fish. Can we still call ourselves a fisher?

Identity may also be defined by identification with a place. Place attachment is the emotional bond between an individual and a place. It has been defined as 'the extent to which an individual values or identifies with a particular environmental setting' (Kyle *et al.*, 2003: 250). There are several aspects to place attachment (Steg, 2018). The first is that the environment satisfies personal and social needs, another is the support of personal and social identity, a third is the positive feelings linked to being in a place (e.g. serenity or wellbeing), and a fourth aspect is the negative feelings of being away from it (e.g. homesickness). Changes in how a marine space is being used may interrupt these place-based bonds, leading to increased opposition to such developments. In the Canadian aquaculture conflict, First Nations feel this link between place and identity strongly:

> It's paramount that we have this strong relationship with the land. It's a land that we consider the Creator bequeathed to us. It's a land where our sacred sites are. It's a land where our first ancestors emanated. From that flowed our membership, our citizenship, our laws, our politics, our economics, our governance, our culture, our potlatch, our spirituality.
>
> (Page, 2007)

After the Exxon Valdez oil spill, indigenous groups argued,

> Try to imagine your church and someone coming in and desecrating it. Look around, this is our church; the land, the birds, the sea, the animals ... this is our church. We have spiritual attachment to these things and we depend on them.
>
> (PR Newswire, 1990)

This connection is felt as keenly in non-indigenous conflicts. For example, in the Cape Wind conflict,

> Those who want the Sound reserved from development hold it as 'place', a living presence that carries the authority of individual and collective memory. It is simple enough to see this as true for the Indians and for those who are focused on the origins of the colonies and the nation, but this power also is the fuel for the passionate work of individuals for whom the Sound represents the core of living memory – of family, of legacy, of a transcendent nature that is made manifest in the water, wind, sand and sun, and all the creatures who abide there.
>
> (North Cairn, 2010)

In aquaculture conflicts in Tasmania, 'Tasmanians have an incredible connection to their marine environment' and 'Many people talk about it being brand Tasmania, but to me it's home and we're going to stand up for it' (The Mercury, 2017).

Threats to the community

While most of us need to be part of a community for life's necessities, we are also social beings who need a sense of belonging. A community brings people together to advocate and support each other in the fight to overcome threats. A community identity may be comprised from social aspects such as the complex network of social histories from which we come, or through the roles we play in the various segments of our social lives (communities of interest). Or it might be a blend of geography, history, economic activity, or even architecture, among other factors (physically located communities). This identity can also be affected by a loss of influence in decision-making, a decrease in population base, or a decline of the economic base through the loss of industry or diminished natural resources. In these ways, changes to how we use marine resources can challenge our community identity.

With colonisation, we saw the destruction of community cultures that might threaten imperial territory. A hangover from colonial times, indigenous peoples are still often excluded from political, economic, social, and cultural life. As such, the concept of a loss of group identity is probably clearest in relation to indigenous cultures. Changing the physical location of indigenous communities or appropriating the resources on which their culture relies (or protecting them so they cannot be used at all) affects indigenous community identity. For First Nations groups involved in aquaculture conflict in British Columbia, when talking about the importance of wild salmon to them, stated 'It's not just a job, not just an opportunity to pass time in recreational mode. It's still very much of who we are. It makes us who we are' (Page, 2007: 623). For an indigenous group affected by the Exxon Valdez oil spill, the loss of subsistence fishing affected their identity 'The Aleutiqq's life of subsistence, they say, is a measure of who they are'. '"Without it, they will lose their identity," says William Simeone, a local anthropologist.' (Sullivan, 1991). However, there are many types of group identity related to the marine environment. Fishing is often valued for its crucial role in the identity of a community. Around the world, we see communities that identify as 'fishing towns' or 'fishing villages', such as Portofino in Italy, Marsaxlokk in Malta, Alvor in Portugal, in which the fishing industry may have declined, but the town continues to trade on its identity for tourism. In Tsitsikamma National Park, with its fishing closures, 'Loss of access to the fisheries resources is synonymous to the loss of cultural identity' (Watts and Faasen, 2009). Group identity is often framed as a 'way of life'. In the Ecuadorian aquaculture conflict 'In these communities, locals are still highly dependent on natural resources, so the environmental effects of mariculture expansion disrupts traditional ways of life' (Crider, 2012).

Linked to identity are cultural knowledges and practices concerning the natural environment, many of which are rapidly disappearing. These include knowledge, skills, practices, and representations developed by communities through interaction with the natural environment. These ways of thinking about the universe are often expressed through language, oral traditions, spirituality, and worldview. They strongly influence values and beliefs, and underlie many social practices and cultural traditions, including traditional ecological wisdom, indigenous knowledge, traditional healing systems, rituals, cosmologies, shamanism, festivals, languages, and visual arts. We are now losing these practices in the marine and coastal environment. In the case of salmon farming in British Columbia, Canada, wild salmon is central to the traditional cultural practices of the First Nations communities in the location. In the Cape Wind conflict, the Aquinnah and Mashpee Wampanoag tribes claimed that the wind turbines proposed for Nantucket Sound would interfere with their spiritual sun greetings.

> The Mashpee and Aquinnah Wampanoag tribes say they need an unobstructed view of Nantucket Sound to carry out spiritual sun greetings and that the waterway's seabed, which was exposed land thousands of years ago, is sacred ancestral land that would be disturbed by building turbines on it.
>
> (Daley, 2010)

They argued that tribal rituals, including dancing and chanting, take place at secret sacred sites around the sound at various times, such as the summer and winter solstices and when an elder passes, and that an unobstructed view of the sun was required for these rituals. For those affected by the Exxon Valdez oil spill,

> Villagers say subsistence living is more than a way to obtain food. It is at the heart of the spiritual and cultural makeup of their lives and was handed down for hundreds of years from their Aleut ancestors. 'If we can't gain back a part of it,' said Elenore McMullen, chief of the nearby Aleut village of Port Graham, 'there's going to be no one to teach it and share it.'
>
> (Enders, 1991)

In many parts of the world, rural and indigenous communities are facing huge cultural, economic, and environmental changes, which contribute to weaken their local knowledge base. This may have an important environmental feedback because of the interconnection between cultural and biological diversity; the loss of local and indigenous knowledge may critically threaten effective conservation of biodiversity, particularly in community-based conservation efforts (Aswani *et al.*, 2018).

Don't govern poorly

Management of the ocean is a complex web of inter-related, intertwined, converging, and competing demands and interests, and this is reflected in the modern governance framework. However, the many institutions that have come into being in recent decades, to arrest the deterioration of the world's ocean and restore it to health, have failed. In fact, a 2014 report by the Global Ocean Commission called *From Decline to Recovery* suggested that existing governance arrangements didn't only fail to stop deterioration of marine resources, but effectively contributed to it (Global Ocean Commission, 2014). The report suggests that a lack of compliance and poor implementation, limited investment in monitoring and enforcement, ineffective co-operation, and limited infrastructure are among the problems that must be addressed if the ocean and the life within are to be saved. As it turns out, failures of participation and trust also contribute to marine and coastal conflict. These link directly to procedural justice as discussed earlier but came through strongly enough in the analysis to warrant their own section.

Failure of engagement and communication

Public participation in natural resource management is increasingly viewed as an important mechanism through which to achieve sustainable development. For this reason, there has been a shift towards greater public involvement in the decision-making process. However, marine and coastal resource management increasingly occurs in turbulent and contentious settings, with little agreement on a variety of aspects from goals to science. Participation can be considered as a tool to achieve a specific end, or as a mechanism for social change. Most often, in the case of ocean and coastal management, it appears to be the former, based on the assumption that once community views have been considered, the policy or project will better reflect their needs and fit into a social and economic reality, and that people will feel a sense of ownership over the policy or project. However, this is clearly not the case.

The International Association for Public Participation has developed a 'Participation Spectrum' to help to define the public's role in any public participation process (www.iap2.org.au). With increasing impact on the decision, this spectrum starts with informing (providing the public with balanced and objective information), moves to consulting (obtaining public feedback), then to involving (working directly with the public throughout a process), to collaborating (to partner with the public in each aspect of the decision), and ends with empowering (placing the final decision in the hands of the public). Most public participation processes focus on the lesser impact end of the spectrum, specifically on informing and consulting. This is a focus of public dissatisfaction underlying marine and coastal conflict.

Projects are often formulated, and operated, without the communication of information among those that it is likely to affect, information that is necessary to prevent any misunderstanding of the projects' aims and objectives. After the Exxon Valdez oil spill, many felt that they had no clue about how the disaster was being dealt with: 'Alaska Natives also demanded the "recovery" of knowledge. They wanted to know what was happening, especially in the remote villages. "I wish someone would just tell me what is going on," said one village resident' (Widener, 2000). In the case of the Brent Spar conflict, although the statutory public consultation had taken place, there had been a lack of communication in other aspects. As noted in a later debrief by Shell of the conflict, 'although in the UK we had carried out statutory public consultation, neither we nor the rest of the industry had explained decommissioning widely enough, early enough' (Shell, 2008). In the South African Tsitsikamma National Park conflict, one academic study noted that, 'The need for more and better communication of information to the communities also surfaced in many of the interviews with key informants as well as in interviews with business owners.' (Watts and Faasen, 2009). In some cases, it is suggested that this lack of communication goes further, that in fact information is purposefully not being provided. In the case of the Supertrawler conflict, it was intimated that information was being purposefully concealed: 'Secrecy surrounding the fishery means we won't know where the vessel is fishing, what it is catching or when dolphins and seals are being killed.' (Hubbard, 2015).

In a conflict around aquaculture in Okehampton Bay, some stakeholders felt that, 'We're not getting enough information, everything is hushed up' (O'Connor and Aird, 2017), suggesting some intent on the behalf of those not communicating, although also potentially reflecting an increase in conspiracy theories.

Being locked out of the decision-making process lies at the heart of a lot of friction in the marine and coastal sphere. Often this displeasure was directed at the government, as opposed to the organisation creating the change. In Canada, in relation to salmon aquaculture, First Nations groups claimed that, 'We have made every attempt to engage Canada in good faith on their PRV Policy and the transfer of Atlantic salmon into our territory but it refused to consult with us' ('Namgis First Nation, 2018). In another aquaculture conflict, in Okehampton Bay in Tasmania, 'the government failed to adequately consult with members of the East Coast community leaving members feeling ignored'. (Wisbey, 2017). Also in Australia, in the Supertrawler conflict,

> Conservation groups have raised the alarm that more large freezer factory trawlers, including one flying a flag of convenience, could be fishing in Australian waters within weeks without any public consultation and despite significant concerns being raised with the Australian Government.
>
> (Hubbard, 2014)

Unhappiness with a lack of consultation also occurred in relation to the implementation of Tsitsikamma MPA 'the communities' desire to be included in discussions about illegal fishing and the park's "no-take" policy on fishing' (Watts and Faasen, 2009) and the disposal of the Brent Spar oil platform, 'Shell and the UK Government acted arrogantly throughout the entire process without considering dialogue as a possibility' (Löfstedt and Renn, 1997).

It may be that there is a good reason for not undertaking public participation. Perhaps the appropriate legal framework to back up such a process does not exist. Public participation can be time-consuming and sometimes expensive and maybe the institutions lack the necessary expertise or other resources. Possibly it is fear, that a negative experience of the process may lead participants to have negative perceptions of the outcome, that is forestalling this activity. Whichever the reason, there are downsides to not consulting the public on matters relating to marine natural resources. Clearly it can create negativity towards the emerging project. Perhaps, more importantly, it means that decision-making fails to take advantage of all available knowledge and expertise.

Even when consultation does take place, many feel that they are not being listened to, making them feel even further disenfranchised. This is the way that First Nations groups felt in the Canadian aquaculture conflict: 'Nobody is listening to us, so if we have to lock doors to get our voices heard, we will.' (Johnson, 2018). In Okehampton Bay, opponents also felt 'fed up with not being listened to' (The Mercury, 2017). Similarly, in the Cape Wind conflict it was suggested that 'People get angry because they feel that they are not listened to' (Schwartz, 2010). Not being heard erodes any sense of personal empowerment. However, consult and you will start to allay stakeholder anxieties or hostility. Listen, and those who feel ignored may re-engage and start to form bonds of trust with those that include them in the process.

Loss of trust

Trust has repeatedly been identified by researchers as an essential ingredient of natural resource management (Schuett *et al.*, 2001, Stern and Coleman, 2015). Researchers have also found that trust held by community members for natural resource decision-making agencies increases public approval of management decisions and minimises resistance to planning efforts. As pointed out by Stern and Coleman (2015: 119):

> Trust can be conceptualized in multiple ways, but most definitions coalesce around the idea that trust is a psychological state in which one actor (the trustor) accepts some form of vulnerability based upon positive expectations of the intentions or behavior of another (the trustee), despite inherent uncertainties in that expectation.

Essentially, trust is a psychological state that indicates a general willingness to accept some degree of risk in the decisions made regarding marine and coastal resources, while distrust suggests the opposite.

A variety of types of distrust are evident in marine and coastal conflicts. For example, in the case of the Exxon Valdez oil spill, trust had previously existed, but was lost in the instant of the spill:

> There was also a loss of trust, which over time had been built between the oil industry and the people of Alaska. Here, it is the oil industry that may need to 'recover' a relationship with the people impaired by the oil spill.
>
> (Widener, 2000: 13)

The situation was similar in the Lyme Bay conflict: 'the events in Lyme Bay, and the Government-imposed ban on scallopers, which has brought all this to a head and destroyed the trust, consultation and co-operation which most fishermen and the less dogmatic conservation bodies were building'. (Western Morning News, 2008). The issue of trust was slightly different in some of the other case studies. In the case of the Brent Spar conflict, the focus of trust problems focused on the actual process of decision-making by both the industry and the government, but it appears that there was at least a little trust,

> Shell's actions did little to instill greater trust among the public. The company was non-univocal, the messages it relayed were at times confusing, and at others arrogant. The UK Government was portrayed by the media as siding with industry, and being both arrogant and stubborn which did not help its public image of trustworthiness.
>
> (Löfstedt and Renn, 1997)

In the Canadian aquaculture conflict, distrust was based on prior history, and trust had not existed for some time, 'Based on failed past ventures in their communities, many First Nations are wary of provincial government involvement in business negotiations with multinational corporate interests in their traditional territories.' (Gerwing and McDaniels, 2006). For some opponents of the salmon aquaculture development in Okehampton Bay, the starting point of the conflict was from a position of distrust,

> The bottom line is that Labor can't be trusted on salmon, they can't be trusted on Okehampton Bay, and they can't be trusted to protect jobs in regional Tasmania because they want to do another election deal with the Greens.
>
> (National Iraqi News Agency, 2017)

Stern and Coleman (2015) suggest there are four types of trust in natural resource governance. The first type of trust is 'dispositional', indicating the predisposition of an individual to trust or distrust an entity in a given context. This may be based on personal history, cultural norms, or cues from the individual's environment. The second type of trust is 'rational' which is based on a calculation of perceived utility of the expected outcome of trusting an entity. This is affected by evaluations of information about the prior performance of the entity and the subsequent predictability of outcomes. The third type is 'affinitive', based entirely on emotions and associated judgements based on assessments of the qualities of the entity. This may result from assumptions of shared values, emotional responses to perceived shared identity, or feelings of social connectedness. Finally, there is 'procedural' trust, which is trust in the procedures or other systems that decrease the vulnerability of the potential truster. This may relate to perceptions of legitimacy and transparency. The authors also point out that it would be unwise for managers to focus exclusively on the development of any one form of trust. Perhaps the most actionable form of trust is procedural trust. On the other hand, wicked problems tend to have typically low levels of dispositional trust and should perhaps be the focus. While no single process is likely to be appropriate in every situation, certain elements may contribute to participants' trust in procedures. While there is no standardised magical set of tools for building trust, understanding what type of trust needs to be addressed may provide a fruitful starting position.

Summary

This chapter has focused on several factors that shape the rules, rights, and effects of human resource use. We have investigated rights to ocean and coastal resources, issues around perceived fairness, potential threats to wellbeing, and concerns regarding poor governance. Each of these factors are split into sub-factors. The first factor, about rights, is split into two components: who has the right to access and who has the right to degrade a resource. The second factor, about fairness, is split into distributive and procedural justice. The third factor, regarding threats to wellbeing, relates to the wellbeing of the individual, but also of the community. Finally, the fourth factor, relating to poor governance, mostly focuses around stakeholder engagement and a loss of trust in the governing institutions.

The results suggest that competition, when it becomes personalised, can lead to conflict. The findings presented in this chapter suggest that in the case studies analysed, competition became personal. The fact that the findings lent themselves to being represented in the form of personal statements (Don't take or alter what is ours (or no-one's); Don't treat us unfairly; Don't threaten our wellbeing; and Don't govern/manage us poorly) highlights this further. Each of these statements are about how competition over resources affects 'us'. Given that the competition is likely to remain, given the drivers identified in the previous chapter, if we want to be able to prevent marine and coastal conflict, we need to consider addressing these 'personalised' factors – ones which can be changed.

References

'Namgis First Nation 2018. 'Namgis First Nation sues to prevent restocking of open-net fish farm at Swanson Island without DFO testing for disease. SK Marketing & Promotions.

Airoldi, L. and Beck, M. W. 2007. Loss, status and trends for coastal marine habitats of Europe. *Oceanography and marine biology*. CRC Press.

Aswani, S., Lemahieu, A., and Sauer, W. H. 2018. Global trends of local ecological knowledge and future implications. *PloS one*, 13, e0195440.

Beck, E. C. 1979. The love canal tragedy. *EPA Journal*, 5, 17.

Beyerl, K., Putz, O., and Breckwoldt, A. 2016. The role of perceptions for community-based marine resource management. *Frontiers in Marine Science*, 3, 238.

Cascao, A. E. and Zeitoun, M. 2010. Power, hegemony and critical hydropolitics. *Transboundary Water Management: Principles and Practice*, 27, 42.
Cifuentes, E. and Frumkin, H. 2007. Environmental injustice: Case studies from the South. *Environmental Research Letters*, 2, 045034.
Crider, M. C. 2012. *The social and environmental effects of shrimp mariculture: Case studies of two coastal villages in Ecuador*. Florida Atlantic University.
Daley, B. 2010. State preservation chief cites wind farm impact: Simon says Native American sites are likely to be harmed. *Boston Globe*, 23 March.
Editorial. 1993. The economic price of pollution. *Atlanta Constitution*, 26 August.
Egan, T. 1989. Alaska's stain: Delicate balance undone. *New York Times*, 1 April.
Electric Utility Week. 2004. Opponents fight for delay of Cape Cod wind farm, even after favorable report. *Electric Utility Week*, 15 November.
Enders, J. 1991. Native village life drastically changed by Exxon Valdez spill. *Associated Press*, 24 March.
Fantin, E. 2017. Fisherman invites Tassal to discuss Okehampton Bay fish farm concerns on protest boat. *Australian Broadcasting Corporation News*, 27 July.
FAO. 2005. *General situation of world fish stocks*. FAO.
Fleming, L., Depledge, M., McDonough, N., White, M., Pahl, S., Austen, M., Goksoyr, A., Solo-Gabriele, H., and Stegeman, J. 2015. *The oceans and human health*. Oxford University Press.
Gallagher, D. 2002. Windmills at sea: Energy entrepreneurs are looking offshore for wind power. *Alternatives Journal*, 22 June.
Gao, Y., Riley, M., and Sadler-Smith, E. 2008. *The cognitive structure of professional identity*. Paper at annual meeting of ANZAM, December 2008.
George, C. and Reed, M. G. 2017. Revealing inadvertent elitism in stakeholder models of environmental governance: Assessing procedural justice in sustainability organizations. *Journal of Environmental Planning and Management*, 60, 158–77.
Gerwing, K. and McDaniels, T. 2006. Listening to the salmon people: Coastal First Nations' objectives regarding salmon aquaculture in British Columbia. *Society and Natural Resources*, 19, 259–73.
Ghosh, A. and Savio Lobo, A. 2017. Bay of Bengal: Depleted fish stocks and huge dead zone signal tipping point. *Guardian*, 31 January.
Gies, E. 2017. Canadian First Nations call for eviction of fish farms. *Hakai Magazine*, 20 October.
Global Ocean Commission 2014. *From decline to recovery: A rescue package for the global ocean*. Global Ocean Commission.
Grotius, H. 1916. *Mare liberum*. Oxford University Press.
Herald Express. 2006. Battle to retain scallop beds. *Herald Express*, 24 May.
Hubbard, R. 2012. *A victory for people power* [Online]. Available: www.stopthetrawler.net/a_victory_for_people_power [Accessed 11 January 2018].
Hubbard, R. 2014. *More foreign factory trawlers* [Online]. Available: www.stopthetrawler.net/more_foreign_factory_trawlers [Accessed 11 January 2018].
Hubbard, R. 2015. *Turnbull government handed opportunity on super trawler* [Online]. Available: www.stopthetrawler.net/turnbull_government_opportunity [Accessed 11 January 2018].
Huon Aquaculture Group Pty Ltd. 2017. *Macquarie harbour submission to EPA* [Online]. Available: https://epa.tas.gov.au/Documents/HUON%20-%20MH%20

Submission%20(Redacted)%20-%20January%202017.pdf [Accessed 23 January 2018].
Johnson, R. 2018. *First Nations face off against open-pen salmon farms in British Columbia* [Online]. Available: www.earthisland.org/journal/index.php/elist/eListRead/First_Nations_face_off_open-pen_fish_farms_british_columbia/ [Accessed 23 March 2018].
Kempton, W., Firestone, J., Lilley, J., Rouleau, T., and Whitaker, P. 2005. The offshore wind power debate: Views from Cape Cod. *Coastal Management*, 33, 119–49.
Knight, W. S. M. 1925. Seraphin de Freitas: Critic of mare liberum. *Transactions Grotius Soceity*, 11, 1.
Kyle, G., Graefe, A., Manning, R., and Bacon, J. 2003. An examination of the relationship between leisure activity involvement and place attachment among hikers along the Appalachian Trail. *Journal of Leisure Research*, 35, 249–73.
Lawrence, R. L., Daniels, S. E., and Stankey, G. H. 1997. Procedural justice and public involvement in natural resource decision making. *Society & Natural Resources*, 10, 577–89.
Lind, E. A., Kanfer, R., and Earley, P. C. 1990. Voice, control, and procedural justice: Instrumental and noninstrumental concerns in fairness judgments. *Journal of Personality and Social Psychology*, 59, 952.
Lind, E. A., Kray, L., and Thompson, L. 1998. The social construction of injustice: Fairness judgments in response to own and others' unfair treatment by authorities. *Organizational Behavior and Human Decision Processes*, 75, 1–22.
Lind, E. A. and Tyler, T. R. 1988. *The social psychology of procedural justice*. Springer Science & Business Media.
Löfstedt, R. E. and Renn, O. 1997. The Brent Spar controversy: An example of risk communication gone wrong. *Risk Analysis*, 17, 131–6.
Manila Bulletin. 2016. The South China Sea: Historical perspectives and contemporary reflections. *Manila Bulletin*, 25 June.
Melosi, M. and Stine, J. 2006. *Echoes from the poisoned well: Global memories of environmental injustice*. Lexington Books.
Mercury. 2017. What some of the protesters said. *Mercury*, 19 June.
Morton, A. 2017. *Marine Harvest vs. First Nations – in court* [Online]. Alexandra Morton. Available: http://alexandramorton.typepad.com/alexandra_morton/page/3/ [Accessed 22 January 2018].
National Iraqi News Agency. 2017. Australia: Labor's half-baked fig leaf to appease the Greens on salmon farming. *National Iraqi News Agency*, 22 February.
Noble, C. 2002. Cape Cod wind farm plan stirs up controversy. *Reuters News*, 31 October.
North Cairn. 2010. The Sound: A precious place. *Cape Cod Times*, 28 March.
O'Connor, T. and Aird, H. 2017. Tasmanian mayor heckled, 'no confidence' in minister, over fish farm pollution fears. *Australian Broadcasting Corporation News*, 5 March.
Ostrom, E., Gardner, R., Walker, J., and Walker, J. 1994. *Rules, games, and common-pool resources*. University of Michigan Press.
Ostrom, E. and Schlager, E. 1996. The formation of property rights. *Rights to nature*, 127–56. Island Press.

Page, J. 2007. Salmon farming in First Nations' territories: A case of environmental injustice on Canada's West Coast. *Local Environment*, 12, 613–26.

Palinkas, L. A., Petterson, J. S., Russell, J., and Downs, M. A. 1993. Community patterns of psychiatric disorders after the Exxon Valdez oil spill. *The American Journal of Psychiatry*, 150, 1517.

PR Newswire. 1990. Impact of the Alaskan oil spill on Native Alaskans: The Untold Story. *PR Newswire*, 14 March.

Schuett, M. A., Selin, S. W., and Carr, D. S. 2001. Making it work: Keys to successful collaboration in natural resource management. *Environmental Management*, 27, 587–93.

Schwartz, E. I. 2010. Lessons from the battle over Cape Wind. *MIT Technology Review*, 1 November.

Selden, J. 1751. *Mare clausum*. Kembe.

Shell. 2008. Brent Spar Dossier. Visual Media Services.

Simon, A. 2012. *Aquaculture and you* [Online]. Stuff.co.nz. Available: www.stuff.co.nz/sport/boating/7376482/Alvah-Simon-Aquaculture-and-you [Accessed 28 November 2017].

Smith, P. D. and McDonough, M. H. 2001. Beyond public participation: Fairness in natural resource decision making. *Society & Natural Resources*, 14, 239–49.

Steg, L. 2018. *Environmental psychology: An introduction*. Wiley-Blackwell.

Stern, M. J. and Coleman, K. J. 2015. The multidimensionality of trust: Applications in collaborative natural resource management. *Society & Natural Resources*, 28, 117–32.

Sullivan, A. 1991. Despite big cleanup many Alaskans feel damage left by oil – U.S. backs natives' demand that Exxon help restore hunting, fishing grounds. *Wall Street Journal*, 5 September.

Thompson Coon, J., Boddy, K., Stein, K., Whear, R., Barton, J., and Depledge, M. H. 2011. Does participating in physical activity in outdoor natural environments have a greater effect on physical and mental wellbeing than physical activity indoors? A systematic review. *Environmental Science & Technology*, 45, 1761–72.

Thompson, C. W., Roe, J., Aspinall, P., Mitchell, R., Clow, A., and Miller, D. 2012. More green space is linked to less stress in deprived communities: Evidence from salivary cortisol patterns. *Landscape and Urban Planning*, 105, 221–9.

Vennells, L. 2008. Fishing rights snatched away. *Western Morning News*, 20 June.

Wainwright, L. 2015. Stakeholder power dynamics in environmental decision making: A case study of the 2012 FV Margiris 'super trawler' fishing dilemma in Australia. Master's thesis, Uppsala Universitet. Available: https://uu.diva-portal.org/smash/get/diva2:818315/FULLTEXT01.pdf [Accessed 22 March 2019].

Walker, G. 2009. Beyond distribution and proximity: Exploring the multiple spatialities of environmental justice. *Antipode*, 41, 614–36.

Watts, S. and Faasen, H. 2009. Community-based conflict resolution strategies for sustainable management of the Tsitsikamma National Park, South Africa. *South African Geographical Journal*, 91, 25–37.

Western Morning News. 2008. Marine conservation must be team effort. *Western Morning News*, 4 November.

Whiting, N. and Salmon, G. 2017. Macquarie Harbour salmon: Huon Aquaculture accuses EPA of playing favourites. *Australian Broadcasting Corporation News*, 15 March.

Widener, P. 2000. *'Recovery' in the media: An Alaska native perspective on the Exxon Valdez Oil spill*. University of New Orleans.
Williams, R. W. 1999. Environmental injustice in America and its politics of scale. *Political Geography*, 18, 49–73.
Wisbey, M. 2017. Students allegedly work on former TasTAFE chief's boats. *Launceston Examiner*, 9 June.
Zacharias, M. 2014. *Marine policy: An introduction to governance and international law of the oceans*. Routledge.

Chapter 7

A new theory of marine and coastal conflict

The findings presented in previous chapters suggest that there may be a relationship between the dependent variable (the focus of the study – conflict) and various independent variables (potential reasons for existence or non-existence of conflict – the drivers and exacerbating factors as identified in Chapters 5 and 6). Currently, however, it is unclear what the relationship is. We want to address the question: why does marine and coastal conflict occur? As such, we need to identify plausible relationships among these sets of concepts, and this is the creation of theory (Corbin and Strauss, 1990).

In this short synthesis chapter, I propose a theory of marine and coastal conflict and lay out an argument to support it. I compare my findings with the existing social and environmental conflict theories detailed in Chapter 3 to determine whether this new theory addresses the gaps identified and offers anything new. I complete this chapter by providing some caveats to this nascent theory.

A new insight?

Based on the findings of previous chapters, I would propose a theory of marine and coastal conflict as follows:

> *Increased competition for ocean and coastal goods and services, driven by a changing social-ecological environment, may be transformed into conflict by political and social factors that shape the rules, rights, and effects of human resource use* (Figure 7.1).

For those of us who make our living from natural resource management or research, this statement of theory may not have a 'wow' factor; it may even come across as common sense. I certainly hope so – theories that we accept tend to be affectionately referred to as common knowledge or common sense (Shoemaker *et al.*, 2003). Where it becomes interesting, and much more relevant to the oceans and coasts, is in what lies beneath the theory, the bits we would put in brackets so to say.

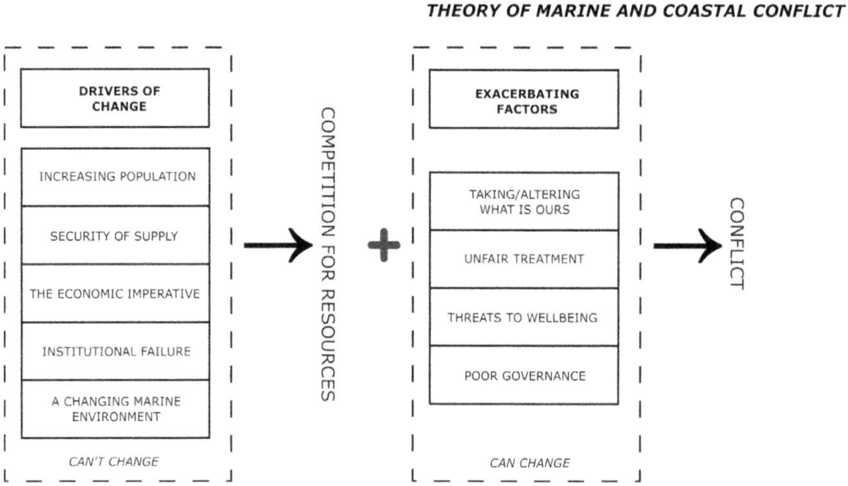

Figure 7.1 A changing social-ecological environment is created by the drivers of change, which leads to competition. Add in exacerbating political and social factors, and this leads to conflict.

Let us start by breaking down this theory into its constituent parts and linking them back to the relevant previous chapters, prior to laying out the argument sitting underneath the theory. The theory comprises several components:

- A changing social-ecological environment
- Increased competition
- Marine and coastal goods and services
- Political and social factors shaping resource use
- Conflict

A changing social-ecological environment refers to the drivers uncovered in Chapter 5: an increasing global population; the need for security of supply; the economic imperative for growth; historical institutional failure in managing the ocean and coastal environment; and a changing marine environment. Increased competition is a result of these drivers. Marine and coastal goods and services are described fully in Chapter 2 and are the 'things' for which there is increased competition. The political and social factors shaping resource use refer to those concerns which are described in Chapter 6: taking/altering what is perceived to be ours; unfair treatment; threats to wellbeing; and continuing poor governance/management. Finally, conflict is that which is defined in Chapter 2, and of which examples are provided in Chapter 4.

The theory itself comprises two key arguments and is important to logically step through each of these arguments to understand the theory. The first argument is that a changing social-ecological environment leads to increased competition for ocean and coastal goods and services. In 2003, the European Union's European Security Strategy identified 'competition for natural resources' as a global challenge. As detailed in Chapter 2, marine and coastal resources play a critical role in the lives of most of the world's population from providing oxygen to absorbing carbon to contributing more than $500 billion to the world's economy. However, as noted in Chapter 5, many factors are driving social-ecological change in the oceans and coasts, leading to increased competition for these resources. We have seen demand for resources grow with an increasing global population, an increasing need for food, water, and energy security, and the imperative for economic growth. This is occurring in tandem with a rapidly changing marine environment in which human impacts are affecting resource availability. Global projections indicate that demand for such resources will continue to increase significantly over the next decades under the pressure of such drivers as these, in addition to changing diets and cultural and technological change (Hoff, 2011). As demand increases, resources will be depleted or degraded leading to increased competition to obtain what remains. Furthermore, increasing demand has outpaced the ability of governing institutions to manage this competition.

As explained in Chapter 5, healthy competition can be a good thing and can deliver a wealth of benefits. It can provide motivation for us to work hard and achieve objectives. It can engender greater commitment to quality as we seek to use this as a lever to beat our competitors. Competition can also serve as a spark to creativity as we seek out all possible avenues to get ahead of the crowd. However, competition also has many cons. According to Sutherland, Woodward, and Maxwell (1961), 'Competition is an impersonal, unconscious, continuous struggle between individuals or groups for satisfaction which, because of their limited supply, all may not have.' Similarly, according to Deutsch (Deutsch, 1977), 'competition implies an opposition in the goals of … interdependent parties such that the probability of goal attainment for one decreases as the probability for the other increases'.

The second is that this competition is exacerbated by political and social factors which shape resource use, and which may tip competition over into the realm of conflict. Pure competition rarely drivers resource conflict. The risk of escalation increases when resource competition becomes linked with other factors such as those identified in Chapter 6. Competition may be exacerbated when we believe that our ability to compete for what we believe is an open-pool resource is being restricted by another/others; when we believe that the process by which resources are distributed, and the resulting allocation of resources to us are unfair and that we don't have the power to influence this; when we believe that our wellbeing or identity is being threatened by another(s); or when we don't agree with how the use of natural resources should be regulated/ managed. Importantly, these beliefs may be imagined or substantive; but if, when and how a conflict arises will depend on the perceptions of those involved. While a competitive situation might exist without any awareness of it by the parties concerned, Boulding (1962: 5) argued that conflict 'is a situation of competition in which the parties are aware of the incompatibility of potential future positions and in which each party wishes to occupy a position that is incompatible with the wishes of the other'. When mechanisms to peacefully address these incompatibilities fail, parties may seek to resolve the matter for themselves and latent conflict will intensify into manifest conflict.

The argument most often used in positing social science theory is that a change in the value of A will lead to a change in the value of B in a linear fashion (Shoemaker *et al.*, 2003). A is the cause, while B is the effect. However, this tends to be untrue. As noted by McGuire (1973: 37), 'Simple a-affects-b hypotheses fail to catch the complexities of parallel processing, bidirectional causality, and reverberating feedback that characterize both cognitive and social organizations.' Indeed, as we can see from the above, the theory presented here to explain marine and coastal conflict is substantially more complex than any sort of simple linear relationship. There are numerous independent variables, split into two groups (drivers and exacerbating factors) which may engender conflict making this a multivariate relationship. Moreover, Chapters 5 and 6 revealed that there are numerous interactions within and between the two groups of independent variables.

Within the larger, overarching and more abstract theory presented here, there are several underlying theoretical statements regarding how the variables interact. Further work will be required to untangle and test these theoretical statements, particularly if we want to be able to predict the future occurrence of marine and coastal conflict. Furthermore, there are questions regarding causal direction. It is unlikely that conflict would cause drivers such as increasing population, the economic imperative and institutional failure to occur, but it may lead to the need for security of supply and changes in the marine environment. Similarly, it is unlikely that conflict would lead us to think someone was taking what was ours or governing us poorly, but it could lead to feelings of unfairness and threats to wellbeing. Further work will also be required to identify the causal direction between variables.

Comparison with existing theories

Does the theory that has just been laid out add anything new to the already diverse range of social and environmental conflict theories? Or does it just repackage what has previously been proposed by others (which may still be useful when viewed through the lens of the unique marine and coastal environment)? In Chapter 3, I identified a number of ways in which the unique characteristics of the ocean and coastal environment had not been dealt with in existing theory. Have these now been addressed?

Two of the three key unique characteristics of the marine and coastal environment, not previous considered in theory, have now been included. This suggests that the theory presented here does offer something new. If we start with the physical ocean and coastal environment, the theory developed here does still focus on the goods or services provided by that environment rather than the physical character itself. The 'causes' of conflict that were identified in Chapters 5 and 6 do not really consider, for example, the fluidity of the marine environment, even in the changing marine environment section. The role of the marine commons, however, plays a substantial role in our understanding of marine and coastal conflict. This is clearest in relation to the exacerbating factor 'don't take or alter what is ours'. Enclosure and exclusion are not the de facto state of affairs when it comes to our understanding of the marine and coastal environment. Rather, we view it as a public resource. The contested nature of this environment also plays a key role in our understanding of conflict in this arena, particularly in terms of the exacerbating factor 'don't threaten our wellbeing' which links directly to individual and group identity in relation to the resources and spaces being discussed.

There are several overlaps with existing theories of conflict, both social and environmental. One clear area of overlap relates to the concept of 'competition' for resources, a key component of the new theory presented here. This could easily be understood as comparable with the Toronto Group's concept of 'resource scarcity'. Of course, when the available resource shrinks, due to depletion or degradation, that is likely to lead to competition. As we have seen in the analysis here, that can be caused by any of the drivers: population growth, the need for security of supply, the economic imperative for growth, an institutional failure to manage the resource, and even a changing environment. The question, of course, is do we compete over ocean and coastal goods and services at any time when scarcity is not an issue? The case studies presented here would suggest not. In every case there is an example of some sort of scarcity, scarcity of fish, scarcity of a view, scarcity of space. Indeed, if there was plenty for everyone, we would not need to compete in the first place.

However, as Homer-Dixon recognised, scarcity alone does not lead to conflict. He believed that scarcity caused intervening conditions which in turn caused conflict. I perceive the situation slightly differently. I suggest that various factors cause human-induced environmental change which leads to competition for resources, that competition alone does not cause conflict, and that there are several synergistic factors that must be combined with competition to produce conflict. This is perhaps more comparable with the Bern-Zurich Group suggestion that environmentally caused conflicts only become violent under certain conditions. I argue that competition only turns to conflict under certain conditions, conditions which are political, social, and cultural. Perhaps the theory I present here aligns most closely with the ideas of the political ecologists, that nature and society are highly interlinked, and that the oceans and coasts comprise an arena of contested entitlements and meanings that are themselves a product of social processes.

Caveats

The construction of theory in the social sciences faces a series of different circumstances from those faced in the physical sciences. The subject matter suffers from massive causal complexity; the topics of interest tend to be generated from normative concerns; and even the most successful models tend to work only in limited contexts whose boundaries are poorly understood. This means that social science theories such as that described here are better understood as models which work to explain a phenomenon in limited settings, rather than holding and applying universally.

While inductive reasoning makes broad generalisations from specific observations, there is often a tendency to hurry towards generalisation. While the case studies used to create this new theory are from around the globe, most occur in developed Western countries. If more case studies were included from developing countries, the findings may be different. Indeed, some might argue that this study includes insufficient empirical data to justify any serious theorising. However, the theory described here is just a first step, a nascent theory, a means by which we can make sense of a jumble of information about marine and coastal conflict, bringing different bits and pieces together in a more useful and coherent form. As such, this theory will require extensive testing and/or cross-validation.

It is most important to note, that I can't currently claim a direct causal link between the findings of Chapters 5 and 6 and the occurrence of conflict. What I can currently claim is that there is a correlation between these findings and conflict and that it appears to be a positive correlation. But correlation does not imply causation. I argue that these various factors intersect. I hypothesise that there is a cause and effect relationship of some sort, based on the logic of the counter-factual conditional (Tacq, 2011), but that further evidence and testing is required to fully establish this relationship. I believe that there is a complex set of conditions, positive and negative, that together are sufficient but not necessary for the outbreak of conflict because other factors, which may not yet have been recognised, could have caused the conflict.

Furthermore, the goal of theory tends to be to explain how something works, but also often to use such explanations to predict things. While much further work is required to develop this theory from an immature to a mature, well-tested theory that can be used for predictions, it still provides us with an understanding of the factors which can contribute to the emergence of marine and coastal conflict. So, the question I would pose at this time is: does it matter that we can't currently claim causality and can't currently make predictions? I would argue that in fact we can start to use the findings of this study to prevent, reduce or resolve conflict. It may be that there is little point focusing on addressing the drivers as these are difficult to influence/cannot be changed. However, we can focus on addressing the exacerbating social and cultural factors.

Summary

Why does marine and coastal conflict occur? Increased competition for ocean and coastal goods and services, which is driven by a changing social-ecological environment, is transformed into conflict by political and social factors that shape the rules, rights, and effects of human resource use. This is my theory of marine and coastal conflict. This clearly overlaps with existing theories of conflict, both social and environmental, particularly in terms of the concept of 'resource scarcity' which is used in a similar way to the way in which I describe the term 'competition'. However, I argue that competition only turns to conflict under certain political, social, and cultural conditions. There are many caveats to this theory, perhaps most importantly the fact that this is a nascent theory, one that requires much further testing and cross-validation. However, it provides us with a 'straw-man' starting point by which we can start to consider conflict within current marine and coastal governance mechanisms. It provides us with a framework to investigate how, and if, these mechanisms can address the 'causes' of conflict – particularly those which may exacerbate the potential for conflict.

References

Boulding, K. E. 1962. *Conflict and defense: A general theory*. Harper.

Corbin, J. M. and Strauss, A. 1990. Grounded theory research: Procedures, canons, and evaluative criteria. *Qualitative Sociology*, 13, 3–21.

Deutsch, M. 1977. *The resolution of conflict: Constructive and destructive processes*. Yale University Press.

Hoff, H. 2011. Understanding the nexus: Background paper for the Bonn 2011 Conference.

McGuire, W. J. 1973. The yin and yang of progress in social psychology: Seven koan. *Journal of Personality and Social Psychology*, 26, 446.

Shoemaker, P. J., Tankard Jr, J. W., and Lasorsa, D. L. 2003. *How to build social science theories*. Sage Publications.
Sutherland, R. L., Woodward, J. L., and Maxwell, M. A. 1961. *Introductory sociology*. Lippincott Company.
Tacq, J. 2011. Causality in qualitative and quantitative research. *Quality & Quantity*, 45, 263–91.

Chapter 8

Conflict and the rule of law

If we consider governance of the marine environment, we are likely to initially think of laws; the legally enforceable rights and duties that govern how we use the oceans and coasts. If we think about conflict in the marine environment, we may ponder the role of law in preventing or resolving disputes. As such, the rule of law is the first marine and coastal governance mechanism that I examine regarding if, and how, it addresses the 'causes' of conflict identified earlier – particularly those factors which may increase the likelihood of conflict.

At the United Nations General Assembly on 'The Rule of Law and Global Challenges', in New York in April 2011, Deputy Secretary-General Asha-Rose Migiro highlighted the importance of legislation in conflict prevention:

> The rule of law plays a critical role in preventing conflict and violence ... Security, access to justice, legal protection for all – these are the building blocks. They aim to help us resolve disputes within and between States through legal, peaceful means, thereby avoiding recourse to violence.
>
> (United Nations, 2011)

Laws have an important role to play in conflict prevention and resolution. But do, or can, laws address those factors that contribute towards conflict over marine and coastal resources? Specifically, do they address the exacerbating factors identified in Chapter 6?

In this chapter, I provide an overview of the different types of law that may influence marine and coastal conflict. I present some examples of where the rule of law has worked to prevent or resolve disputes. Finally, and most importantly, I identify if, and how, law can address the four exacerbating factors: our rights (or lack of) to common resources, our needs for social justice, threats to our wellbeing and poor governance.

International treaties, domestic legislation, and customary law

While human use of the marine environment began approximately 164,000 years ago, and the Lex maritima (common law – general maritime law) can be traced as far back as the eighth or ninth century BC (Tetley, 1994), it was only with the rise of secular states in Western Europe in the seventeenth century that ocean governance and legal systems began to emerge (Zacharias, 2014). Marine law has been differentiated from land law as far back as Roman times, often because it tended to be international in nature. Indeed, over 150 international conventions of relevance to the marine environment have been adopted (ibid.).

Perhaps the most well-known piece of international maritime legislation is that of the United Nations Conventions on the Law of the Sea (UNCLOS; introduced in Chapter 2). This law was created because by the mid-twentieth century, human use of the oceans had changed significantly with increased offshore oil and gas development, increased fishing, increased container-based shipping and seabed mining, leading to several nations unilaterally increasing their state jurisdiction beyond the (previously recognised) 12 nautical mile territorial seas. Unsurprisingly, this led to friction between coastal states. Between 1958 and 1972, three United Nations Conferences on the Law of the Sea were held to address this issue, leading to the development of perhaps the most complex and comprehensive international agreement, the UNCLOS treaty, which was adopted and signed in 1982 and came into force in 1994. Due to its complexity, UNCLOS took a long time to be ratified (formally consented to) and several countries have signed the agreement, but not ratified it, including the United States. Significant issues covered in the treaty included navigation, transit regimes, exclusive economic zones (EEZs), continental shelf jurisdiction, deep seabed mining, protection of the marine environment, scientific research, and settlement of disputes. UNCLOS contributes to the prevention of marine and coastal conflicts by explicitly detailing who has rights of access to specific geographical areas.

UNCLOS also has an in-built dispute mechanism. Part XV of the Convention lays down a comprehensive system for the settlement of disputes over the interpretation and application of the Convention. It requires State Parties to settle their disputes concerning the interpretation or application of the Convention by peaceful means as indicated in the Charter of the United Nations. However, if the disputing parties are unable to reach a settlement by peaceful means of their own choice, they are obliged to resort to the compulsory dispute settlement procedures entailing binding decisions, subject to limitations and exceptions contained in the Convention (International Tribunal for the Law of the Sea, n.d.). UNCLOS contains both negotiation and arbitration approaches to dispute resolution.

Arbitration approaches include submission to: The International Tribunal for the Law of the Sea (including the Seabed Disputes Chamber), the International Court of Justice, Ad hoc arbitration, and special arbitration tribunals. If the parties involved do not accept the same settlement procedure, the dispute may be submitted only to arbitration.

Although UNCLOS is the most recognised international treaty in the marine space, there are several other international treaties which may also influence the occurrence of marine and coastal conflict. For example, the International Convention for the Prevention of Pollution from Ships (MARPOL) is the main international convention aimed at the prevention of pollution from ships caused by operational or accidental causes. It was adopted at the International Maritime Organization (IMO) in 1973. MARPOL is one example of a treaty which can prevent conflict by addressing the issue of subtraction discussed in Chapter 6. Another example is the Straddling Fish Stocks Agreement (formally, the Agreement for the Implementation of the Provisions of the United Nations Convention on the Law of the Sea of 10 December 1982 relating to the Conservation and Management of Straddling Fish Stocks and Highly Migratory Fish Stocks). This agreement is a multilateral treaty created by the United Nations to enhance the co-operative management of fisheries resources that span wide areas and are of economic and environmental concern to more than one nation. These international treaties are just two examples of many.

In the 1970s, the regional seas became the subject of international conventions. Regional seas are those marine areas bordered by several neighbouring countries. There are 18 Regional Seas Conventions and associated action plans. The conventions include: the OSPAR Convention for the North-Eastern Atlantic, the Helsinki Convention (HELCOM) for the Baltic Sea, the Barcelona Convention and associated Protocols for the Mediterranean Sea, and the Bucharest Convention for the Black Sea, among others. The conventions were created to tackle shared environmental issues through joint coordinated activities. Most conventions have added protocols, legal agreements addressing specific issues such as protected areas or land-based pollution. The coordinated nature of the conventions affords the potential to prevent or at least reduce conflict between states. Of course, such international legislation is useful in the case of maritime disputes between states. However, it is not as helpful for addressing the majority of conflicts discussed in this book – small-scale, non-violent conflicts which occur on a national, if not regional or even local scale.

Efforts to resolve national level environmental problems as well as competition between many uses and users have led to an extensive patchwork of legislation. So what forms of legislation might relate to the prevention of small-scale non-violent conflicts? Any legislation which attempts to give usage rights to a group or individual is attempting to prevent conflict by awarding ownership of that resource (in addition to transforming the natural resources into a means for domestic or foreign accumulation). Some legislative tools and techniques, used to manage fish stocks, were discussed in Chapter 6. These might include input controls which are designed to limit either the number of people fishing or the efficiency of fishing or they may include restrictions on gear types, number and size of vessels, the area fished, the time fished, or the number of fishers. These regulations can also include output controls which directly limit catch and can be used to set limits for an entire fleet or for a fishery (e.g. by setting a total allowable catch), to set limits for specific vessels (individual vessel quotas), or to set limits for owners or operators (individual fishing quotas). By regulating costs and benefits among user groups, there is an attempt to neutralise conflict. Similarly, by allocating seabed use licences to activities such as oil and gas extraction, renewable energy extraction, seabed mining, or aquaculture, we are privatising that resource, and reducing conflict created by other users wanting to access the same resource. Often, however, legislation of this sort is not very successful at preventing or reducing potential conflict, and legislation then comes into play as part of dispute resolution through arbitration. These different forms of national legislation may also come into conflict with each other and this may lead to further conflict. An example of this is the conflict experienced between legal requirements to manage activities in the marine environment to achieve good environmental status under the European Marine Strategy Framework Directive and other legislation to enable social and economic growth.

Customary law, a traditional common rule or practice which is treated as a legal requirement, is also common practice in some countries around the world, mostly in areas with indigenous populations, such as the small island developing states of the South Pacific. Customary laws are central to the very identity of indigenous peoples and local communities, defining rights and responsibilities of the community members, and forming an intrinsic part of social and economic systems. One particularly relevant aspect of customary law is that of customary marine tenure (CMT). CMT operates under a diversity of different rules and practices across various societies, but the fundamental characteristics of CMT systems include defined geographical areas, controlled access, self-monitoring, and the enforcement of rules and regulations by local people and their traditional leaders (Bennett, 2012). Oftentimes, there is a focus on equal opportunities of access or equitable distributions of income. Where conflicts do arise, local courts or community leaders often act as arbitrators. However, the methods of arbitration differ greatly across cultures. In Somalia, for example, the Xeer system facilitates reconciliation and restorative justice (Abdile, 2012). The Navajo people use a community-based dispute resolution ceremony which integrates the wisdom, skills, and perspectives of various participants to reach noncoercive settlements (Brown, 1999).

It should be noted that CMT does not always resolve conflicts, in fact, conflicts can arise regarding CMT such as those described by Dedy Supriadi Adhuri in *Selling the Sea, Fishing for Power* (Adhuri, 2013). Furthermore, CMT is not considered purely as an instrument to deal with conflict, but is a political asset, used by local elites to control power in the community (ibid.).

Using law to prevent or resolve conflict

While it is possible to go online and find the list of cases tried in, for example, the International Tribunal for the Law of the Sea, including the 25 (as of February 2019) contentious cases, or indeed any other national court, it is not as easy to find other examples of how legislation has been used to resolve or prevent conflicts. However, a few examples are provided here, covering international law, national law, and customary law.

The 'Volga Case' between Russia and Australia

In early 2002, a long-line fishing vessel under a Russian flag – the Volga, was detained by Australian authorities a few hundred metres outside the EEZ of Australia's Heard and McDonald Islands in the Southern Ocean. The vessel was believed to be engaged in illegal fishing, one of the main maritime security issues for Australia. This incident led to litigation in Australian and international courts.

As explained by Karim (2011), after being apprehended, the Russian vessel was escorted to the port of Fremantle in Western Australia, at which point the master and three other members of the crew were detained. A seizure notice was served on the master stating that the vessel with all equipment and catch would be forfeited if a notice of claim was not submitted to the Australian Fisheries Management Authority by the owner or possessor. After this, the three crew members were charged with illegal fishing in the Australian Fisheries Zone. The Australian authorities sold the catch of the Volga and held the money in trust. Following this, the owner of the Volga started proceedings in the Federal Court of Australia to stop the forfeiture of the vessel. The highest court of Australia confirmed the forfeiture of the vessel, and the three crew members pleaded guilty. It is within this context that the Russian Federation commenced proceedings against Australia in the International Tribunal for the Law of the Sea, for the prompt release of the vessel and its crew.

The Russian application stated that the vessel was on the high seas at the time of boarding, that no government vessel had ordered the boat to stop while the vessel was in internal waters, and that the bond set by Australia was unreasonable. Accordingly, the applicant argued that Australia was in breach of articles 87(1)(a) and 111 of UNCLOS when it boarded the vessel (section 87 grants power for surveillance and enforcement and 111 relates to 'hot pursuit' of a vessel).

In this example, legislation was used to underpin arbitration, as opposed to being used in advance to prevent conflict. The Tribunal unanimously found that it had jurisdiction to hear the case made by the Russian Federation. Moreover, the tribunal found by 19 votes to 2 for the allegation made by the Russian Federation that Australia had not complied with the provisions of UNCLOS and decided that Australia should promptly release the vessel upon the posting of a bond (White and Trevisanut, 2002). The Tribunal decided that the bond should be A$1,920,000 which was equal to the value of the vessel including all equipment, fuel, and net.

Galápagos Marine Reserve and the Special Law for Galápagos 1998

Between 1992 and 1995, the long-term sustainable management of the Galápagos Marine Reserve was threatened by severe multi-stakeholder conflicts at both local and national level over the management and exploitation of the marine resources within the area. In 1995 there were violent protests at the Galápagos National Park Service headquarters and the Charles Darwin Research Station (Heylings and Cruz, 1998).

The Galápagos Marine Reserve, recently granted UNESCO World Heritage Site status, is the largest marine reserve in a developing country and the second largest reserve in the world. Covering an area of around 133,000 km^2, the reserve encompasses marine habitats such as rocky seabed, vertical rock faces, sandy beaches, mangrove swamps, and coral reefs, the area hosts a diverse range of animal life. Located at the confluence of three major ocean currents, mixing warm and cold and nutrient rich waters, the reserve has an exceptional marine biodiversity and primary production that supports a very high biomass of pelagic species (Moity, 2018). However, the Reserve faces many threats, particularly overfishing and illegal fishing, including illegal sea-cucumber and shark-fin fisheries. Conflicts arose due to differing interests between Reserve users, for example between local fishermen and tour operators; between conservationists (local and international) and fishermen; and between Galápagos stakeholders and industrial fishing groups from the Ecuador mainland (Heylings and Cruz, 1998).

The Galápagos Marine Reserve was created in 1986 by special decree, however, due to legislative complications (which are beyond the scope of this case example), enforcement was difficult. In 1996 the National Protected Areas Authority unilaterally declared it a Biological Reserve, bringing it under national protected area legislation. Unsurprisingly, this was contested by the Sub-secretary for Fisheries and the industrial fishing sector. Regardless, the declaration of a Biological Reserve provided the mandate to coordinate revision of the management plan. At the same time, the Special Regime Law for the Conservation and Sustainable Development for Galápagos was being drafted, opening up the possibility that the change in management desired by the local stakeholders could be achieved through the Special Law process (Heylings and Cruz, 1998).

The enactment of the Special Law in 1998 marked an historic milestone for fisheries for two primary reasons: (1) it created the Galápagos Marine Reserve, with an extension of 40 miles measured from the outer coastlines; and (2) it restricted fishing activities within the reserve to local fishermen who were affiliated with a legally constituted fishing co-operative. This did, of course, lead to further conflict with those fishers no longer able to fish within the area! Key to the Special Law was the concept of co-management (see Chapter 10) and since its inception, this process has sought to resolve conflicts by creating consensus for the management of the Galápagos Marine Reserve. It has been argued that while co-management faced a few serious challenges, it managed to make the necessary decisions for the conservation and sustainable development of the reserve (Llerena *et al.*, 2017).

Customary law and the Locally-Managed Marine Area (LMMA) Network, Fiji

The Republic of Fiji is an island country in the South Pacific Ocean; an archipelago of more than 330 islands and more than 500 islets, covering an area of about 18,300 km^2. It is an area of high biodiversity with a variety of tropical marine habitats including oceanic reefs, near shore fringing reefs, mangrove forests, sea grass beds, lagoons, and estuaries. The Fijian ocean and coastal environment also provides various benefits from marine ecosystem services, including through human activities such as: subsistence fishing valued at US$29.5 million per year to Fijian households, inshore commercial fishing, offshore commercial tuna fishing valued at US$10.1 million in 2013, sand and aggregate mining, deep sea mineral extraction, and tourism valued at US$574 million in 2013 (Gonzalez and Ram-Bidesi, 2015).

National legislation in many Pacific island countries recognises and protects indigenous land tenure, with most land held under customary communal title. Indeed, customary governance systems remain the primary mechanism for regulating the use of marine resources in many contemporary Pacific societies, such as Fiji, partly because respect for customary law and institutions is an integral feature of most rural communities (Clarke and Jupiter, 2010). While the population of Fiji remained low, and adherence to customary law high, traditional conservation mechanisms worked well. However, with urbanisation and globalisation they began to fall into disuse, or became ineffective, and began to put marine areas under threat (Techera,

2008). As pressure on the marine environment has led to further and further deterioration, there has been a large shift back towards customary law and community-based management of the ocean and coastal environment in these states. One such example is the network of LMMAs that has resulted in 135 of Fiji's marine *iQoliqolis* (customary fishing areas) being managed and 456 fishing reserves covering just over 1,000 km^2.

The LMMA process involves the local community developing a marine Management Plan for the Project Site, as well as implementing, monitoring and enforcing the management plan (Techera, 2008). The management plan usually incorporates traditional knowledge and conservation practices combined with modern scientific knowledge and monitoring methods. Tools used include limited harvesting restrictions for particular species and behavioural controls such as specific gear or equipment bans. These tools are often implemented using customary legal mechanisms.

Several examples of LMMAs are provided on the LMMA network website (lmmanetwork.org). However, one example of customary law helping to prevent conflict describes the relationship between a beach resort and an adjacent fishing village. On Fiji's big island of Viti Levu, approximately 130 km west of the capital Suva, exists the Hideaway resort and Tagaqe Village in Korolevu-i-wai district. When the resort was bought by new owners, they wanted to use their beachfront area for tourism. However, this area was part of a 6 km^2 *qoliqoli* (a traditional fishing ground) where species such as mangrove crabs, clams, octopus, lobster, sea urchin, and reef fishes were caught by the adjacent fishing village. The new owners visited the Chief of Tagaqe Village, Ratu Timoci Batireréga, and asked his permission to use the waters for tourism. The area in front of the Hideaway has since been made '*tabu*' meaning that there is a ban on fishing in this area. Together, they have contacted the network to create a project to address increasing threats caused by destructive fishing methods, coral harvesting, pollution, and coral trampling.

2012 Fisheries Law, Chile

It is also worth noting that the creation of legislation can create conflict between different user groups. In Chile, the line between artisanal and commercial fishers is not clear, particularly for the largest artisanal boats which can require crews of up to ten men (Nelsen, 2013). For years, animosity has bubbled just below the surface between the two groups.

Chile had adopted its first law to regulate fisheries in 1991, but the El Niño effects on climatic conditions in 1997, and the subsequent impact on the livelihoods of fishing communities, created a need to revise these regulations. In 1999 Chile adopted Total Allowable Catch for its main species, then established individual transferable quotas to divide the catch among companies and individuals engaged in fishing. A ten-year harvest limit plan was established in law in 2002. However, there were several weaknesses in the law and it was decided to improve the 2002 law when it expired in 2012 (EFD, 2013).

As Nelsen (2013) explains, in 2011, the Minister of the Economy brought together the various unions to try to settle long-standing disputes, but common ground was not identified, in fact, the negotiations exposed further divide. As such, various groups were in conflict regarding the newly proposed 2012 fisheries law. One component of the artisanal fisheries signed an agreement to co-operate with the industrial sector. Another solidified behind a media campaign. This campaign was funded by a Norwegian company who lobbied for a change from the existing quota system to auctions which would enable it to gain market share. Furthermore, shortly after the law was passed in December 2012, several senators filed a lawsuit challenging the law, claiming that indigenous communities had not been consulted. The law remains in place.

Law and the causes of conflict

The examples provided above make clear that legislation can be used to prevent conflict from happening in the first place and to resolve it (in terms of arbitration) when it does happen. Now that we have a framework by which to more fully understand if we are addressing the causes of conflict, or at least the exacerbating factors, the question is: does legislation deal with those aspects that really can heighten conflict in the first place, and if so, how? Analyses would indicate that legislation has much to offer in addressing some of these issues (Table 8.1).

Don't take/alter what is ours

It could be argued that without legal protection, at least in the form of who has rights to what, conflicts among competing groups are inevitable (Berkes, 2005). International treaties and domestic legislation help to address this by determining the rights of users to access and exploit ocean and coastal resources. Indeed, this is the factor which international treaties and legislation are most directly able to address. As described earlier in this chapter, UNCLOS is one example of this, fisheries legislation and seabed rights are others. However, many countries appear to have an unnecessarily complex marine legislation and administrative framework, responding to a whole suite of international, regional (e.g. European) and national laws which control many different sectors such as fisheries, energy and conservation. On typing 'Who owns the oceans' into Google (in January 2019), approximately 4 million results appear; given the complexity of international treaties and domestic legislation, it is no wonder we (citizens) are confused about who has rights to which marine and coastal resources.

Table 8.1 Overview of if, and how, law addresses exacerbating causes of conflict

Exacerbating factor	*Sub-factor*	*Does law address (yes/no)?*	*Reason*
Don't take/alter what is ours	Exclusion	Yes	International treaties and domestic legislation can determine and clarify the rights of users to access and exploit ocean and coastal resources, e.g. UNCLOS, Fisheries Acts.
	Subtraction	Yes	International treaties and domestic legislation can address subtraction by limiting levels of resource exploitation and addressing degradation caused by e.g. pollution.
Don't treat us unfairly	Distributional justice	No	International treaties and domestic legislation cannot address social comparison.
	Procedural justice	Yes	Processes for procedural justice can be incorporated into law but this does not happen in all instances.
Don't threaten our wellbeing	Individual	Yes and no	International treaties and domestic legislation do not prevent loss of subsistence or livelihood, or effects on identity. They do legislate for health and safety issues.
	Community	Yes and no	International treaties and domestic legislation protect culture and heritage (e.g. indigenous title), but not community identity more generally.
Don't govern/manage us poorly	Engagement and communication failure	Yes	Stakeholder engagement is part of the process of enacting legislation and can be part of the legislated process for decision-making.
	Loss of trust	No	International treaties and domestic legislation cannot legislate for trust.

International treaties and domestic legislation can also address subtraction. First, they can limit levels of resource exploitation. This is evident in the case of fisheries as described earlier in this chapter. However, there are many types of resource extraction which are not limited (surprisingly) by legislation such as oil and gas, sand and gravel, and mineral resources. Second, laws can address degradation in cases such as pollution (although it should be noted that once pollution spreads, it is difficult to address with treaties or legislation). There are many examples of ways in which pollution is addressed in law. The Barcelona Convention for the Protection of the Marine Environment and the Coastal Region of the Mediterranean was adopted in 1995 and has had significant pollution control successes. For example, for the eight Mediterranean states that are members of the European Union (EU), on average, 58 per cent of the wastewater receives secondary treatment and 66 per cent of that receives tertiary treatment. Moreover, discharges of six major industrial pollutants (Mercury, Cadmium, Lead, Zinc, Chrome, and PAHs (polycyclic aromatic hydrocarbons)) have all been significantly reduced (Smithsonian Institution, n.d.). In the Philippines, Muntinlupa City was the first local government in Metro Manila to implement a ban on plastic bags in 2011 (officially doing so via Ordinance 10–109). Ordinance 10–109 has had a domino effect, with other cities following suit with their own plastic bag bans. There is now a push for a nation-wide plastic bag ban (Legaspi, 2018). Clearly, law has much to offer in terms of clarifying who has rights to which resource, as well as who has rights to alter that resource. Elucidation of this information is likely to reduce the potential for conflict.

Don't treat us unfairly

Substantive distributive justice relates to the fairness of outcomes. Treaties and legislation are often designed to ensure the outcome of a process. Regardless of the specific outcomes, if public participation procedures are perceived as fundamentally fair, then the outcome, even if unfavourable, is likely to be perceived as acceptable.

According to John Rawls in *A Theory of Justice* (2009), pure procedural justice is obtained when there is no independent criterion for the right result: instead there is a correct or fair procedure such that the outcome is likewise correct or fair, whatever it is, provided that the procedure has been properly followed. Processes for procedural justice can be incorporated into laws but this does not always happen. It has been argued, however, that the evolution of domestic and international law is trending in this direction thanks to its increasing focus on the public participation process (Parlow, 2018). This links directly to the fourth exacerbating factor regarding poor governance and as such will be discussed further there.

Furthermore, as identified in Chapter 6, a person may seek judicial review of an administrative decision on the basis that procedural fairness has not been observed. In 2014 a community group called 'Friends of Loch Etive' took Argyll and Bute Council (a local planning authority in Scotland, United Kingdom) to judicial review following a decision by the Council granting planning permission for a rainbow trout farm in Loch Etive. Judicial review is the power of the courts of a country to examine the actions of the legislative, executive, and administrative arms of the government and to determine whether such actions are consistent with the constitution. Essentially, it rules on whether the decision-maker has followed due process. In the Loch Etive case, the petition by the community group was refused.

Don't threaten our wellbeing

In terms of individual wellbeing, it is unlikely that we would legislate to prevent loss of subsistence or livelihood, or even effects on identity. However, we do legislate to ensure health and safety. There are many international standards that are reflected in domestic law, for instance, the International Convention for Safety of Life at Sea, the Convention on the International Regulations for Preventing Collisions at Sea, and MARPOL. In Australia, for example, these standards are put into effect through the Navigation Act 2012. However, Australia also has other laws concerning marine and coastal health and safety, enshrined in the Marine Safety (Domestic Commercial Vessel) National Law Act 2012.

Similarly, in regard to community wellbeing, it would be unlikely that we would legislate to protect general community identities, but we do legislate to protect culture and heritage. Aboriginal rights and title are being fought for in several countries around the world, in part to protect aboriginal culture and heritage. In British Columbia, Canada, there are two titles recognised in law – Aboriginal Title and Crown Title. The Canadian legal system recognises Aboriginal title as *sui generis*, or the unique collective right to the use of and jurisdiction over a group's ancestral territories. In Malaysia, Native Customary Rights and Native Customary Land are provided for under various pieces of legislation including the National Land Code 1965, the Sarawak Land Code 1957, and the Customary Tenure Enactment. Aboriginal titles also exist in New Zealand, Australia, Belize, Botswana, Papua New Guineau, and the United States, although they have continued to be a source of contention. It is clear that legislation has a role to play, at least in part, to address this exacerbating factor.

Don't govern/manage us poorly

Legal instruments can fully address this exacerbating factor in various ways. For a start, stakeholder engagement is part of the process of enacting legislation. Citizen engagement at this stage might include dialogue allowing the identification of issues and solutions, a poll of citizens' views on a specific bill, input on a specific issue, input on the text to be included within a specific bill or dialogue on the implementation of government policy, and/or input on the implementation of specific government policy or as part of an investigation of a specific case. Almost all legislation has some form of public engagement before being subjected to the ultimate form of public engagement – parliamentary approval. However, whether this latter component is acceptable relates to the second aspect of this factor – trust – whether we trust our parliamentary representatives to represent us. Unfortunately, international treaties and domestic legislation cannot address the issue of trust.

Stakeholder engagement can also be part of the legislated process for decision-making. In many countries around the world there is a requirement for engagement with statutory stakeholders during the early stages of development projects. Such stakeholders may include marine management and safety authorities, local councils, natural heritage organisations, fishers' associations, marine parks authorities, etc. However, this is not always required. As we have seen from the conflict case studies, the statutory stakeholders are often a small sub-group and not representative of all marine and coastal stakeholders. This means that many stakeholders are left without a voice regarding how the legislation is enacted.

Summary

The rule of law underpins the way society is governed. Everyone, including citizens and the government, is bound by and entitled to the benefit of laws. As such, laws can have an important role to play in conflict prevention and resolution. There are various types of law that may be of importance in preventing or resolving marine and coastal conflict. These include: international treaties, regional conventions, domestic legislation, and customary law. While we know that the rule of law plays an important part in the process, it is often difficult to find specific examples to detail this. Regardless, three examples have been provided in this chapter detailing how law has helped to address marine and coastal conflict, at the international, national, and local levels.

Laws can determine the rights of users to access and exploit ocean and coastal resources, e.g. UNCLOS, Fisheries Acts. They can also address subtraction limiting e.g. levels of resource exploitation or pollution. Processes for procedural justice (and potentially as a result distributional justice) can be incorporated into law, although this does not always happen. Law cannot prevent loss of subsistence or livelihood, or effects on identity, but it can ensure health and safety. Law can protect also protect culture and heritage (e.g. indigenous title), but not community identity in general. Last, but not least, stakeholder engagement is part of the process of enacting legislation and can be part of the legislated process for decision-making.

Clearly law has a role to play in addressing marine and coastal conflict. As we have seen, however, there are many aspects of the exacerbating factors that are not, or cannot, be addressed by law, whether at the international, regional, domestic, or local level. Laws are shaped by, and interpreted according to, context. That context can change, leading to a situation where laws cause, rather than resolve marine and coastal conflict. This leads me to believe that we cannot rely on law alone to prevent, reduce, or resolve such conflict.

References

Abdile, M. 2012. Customary dispute resolution in Somalia. *African Conflict and Peacebuilding Review*, 2, 87–110.

Adhuri, D. S. 2013. *Selling the sea, fishing for power: A study of conflict over marine tenure in Kei Islands, Eastern Indonesia*. ANU E Press.

Bennett, G. 2012. *Customary marine tenure and contemporary resource management in Solomon Islands*. 12th International Coral Reef Symposium, Cairns, Australia, 9–13 July.

Berkes, F. 2005. Commons theory for marine resource management in a complex world. *Senri Ethnological Studies*, 67, 13–31.

Brown, H. L. 1999. The Navajo nation's peacemaker division: An integrated, community-based dispute resolution forum. *American Indian Law Review*, 24, 297.

Clarke, P. and Jupiter, S. D. 2010. Law, custom and community-based natural resource management in Kubulau District (Fiji). *Environmental Conservation*, 37, 98–106.

EFD. 2013. *Sustainable Fisheries Law Promotes Reliable Fishing in Chile* [Online]. Available: http://efdinitiative.org/our-work/policy-interactions/sustainable-fisheries-law-promotes-reliable-fishing-chile [Accessed 15 January 2019].

Gonzalez, R. and Ram-Bidesi, V. 2015. *National marine ecosystem service valuation summary report: Fiji*. Marine and Coastal Biodiversity Management in Pacific Island Countries (MACBIO).

Heylings, P. and Cruz, F. 1998. *Common property, conflict and participatory management in the Galapagos Islands*. Proceedings of the International Association for the Study of Common Property (IASCP) Conference 1998.

International Tribunal for the Law of the Sea. n.d. *The Tribunal* [Online]. Available: www.itlos.org/en/the-tribunal/ [Accessed 12 February 2019].

Karim, M. 2011. Conflicts over protection of marine living resources: The Volga case revisited. *Goettingen Journal of International Law*, 3, 101.

Legaspi, A. 2018. Senators to push for single-use plastic ban. *GMA News*, 7 June.

Llerena, E., Quisingo, T., and Maldonado, R. 2017. Analysis of agreements reached in the Participatory Management Board 2010–2015. *Galapagos Report 2015–2016*. Puerto Ayora, Galapagos, Ecuador: GPND, GCREG, CDF and GC.

Moity, N. 2018. Evaluation of no-take zones in the Galapagos Marine Reserve, Zoning Plan 2000. *Frontiers in Marine Science*, 5, 244.

Nelsen, A. 2013. *Chile: Artisan Fishermen vs. Industrial Fishermen* [Online]. Available: http://pulitzercenter.org/reporting/chile-artisan-fishermen-vs-industrial-fishermen [Accessed 15 January 2019].

Parlow, A. L. 2018. Toward distributive justice in offshore natural resources development: Iceland and Norway in the Jan Mayen. *Ocean & Coastal Law Journal*, 23, 147.

Rawls, J. 2009. *A theory of justice*. Harvard University Press.

Smithsonian Institution. n.d. *Reducing pollution in the Mediterranean Sea* [Online]. Available: https://ocean.si.edu/conservation/fishing/how-laws-around-world-help-improve-ocean-health [Accessed 13 February 2019].

Techera, E. J. 2008. Supporting the role of customary law in community-based conservation. *Journal of the Australasian Law Teachers Association* 279–92.

Tetley, W. 1994. The general maritime law: The Lex Maritima. *Syracuse Journal International Law & Commerce*, 20, 105.

United Nations. 2011. *Rule of law critical in preventing conflict, providing platform for sustainable economic development, Deputy Secretary-General tells General Assembly* [Online]. Available: www.un.org/press/en/2011/dsgsm546.doc.htm [Accessed 18 December 2018].

White, M. and Trevisanut, S. 2002. The Volga case: The judgment. *Maritime Studies*, 2002.

Zacharias, M. 2014. *Marine policy: An introduction to governance and international law of the oceans*. Routledge.

Chapter 9

Spatial planning for conflict prevention

The expansion of offshore activities and the increasing need to meet international and national commitments to biodiversity conservation, leading to increasing conflict between users of the oceans and coasts, has driven interest in marine spatial planning (MSP) as a tool for sea use management. This is the second governance mechanism I will test against the exacerbating factors framework used in Chapter 8.

Defined by the Intergovernmental Oceanographic Commission Marine Spatial Planning Programme as 'a public process of analyzing and allocating the spatial and temporal distribution of human activities in marine areas to achieve ecological, economic and social objectives that have been specified through a political process' (IOC-UNESCO, n.d.), I ask if, and how, MSP can address those factors which may fan the flames of marine and coastal conflict.

In this chapter, I provide an overview of MSP, what it is, and what it isn't. I present some examples of MSP processes that have been undertaken, identifying how they dealt with issues of conflict. Finally, and vitally, I identify if, and how, MSP can address the four exacerbating factors: our rights (or lack of) to common resources, our needs for social justice, threats to our wellbeing, and poor governance.

Marine spatial planning

Many places already designate or zone marine space for human activities such as maritime transportation, oil and gas development, offshore energy, seafood production, and waste disposal. However, this tends to be done on a sector-by-sector basis without much consideration of effects on other human activities or the marine environment. MSP provides a mechanism for looking at the bigger picture; managing existing and potential conflicting uses and the cumulative effects of human activities and marine protection.

MSP comprises several common denominators (Zaucha and Gee, 2019). First, it considers all four dimensions of the marine environment, the sea surface, the water column, the sea bottom, and the subsoil beneath it. It encompasses both space and time. MSP concerns human activity and its consequences, and it incorporates individual human preferences in relation to marine space by a process of public choice. It is integrated. It refers to the sea as a functional ecosystem. It influences market processes at sea. MSP requires transnational coordination within sea basins and it requires coordination and connection with spatial planning on land. It is conducted in a continuous and adaptive manner encompassing monitoring and evaluation. Finally, it employs best available research and information. However, as Zaucha and Gee (2019) also note, the concept is still contentious and there is debate over several aspects, for example the degree of generality/specificity, relations between planning and management, and the scope and methods of engaging stakeholders in the process.

There are also many things MSP is not. MSP is not ocean zoning, although zoning can be an integral part of it. It does not replace sectoral planning and programming and should not be confused with licensing and permitting. It is not a one-time solution. It needs multiple iterations to achieve sustainability. It needs to be flexible and adaptive in what is a dynamic and constantly changing environment. Finally, despite what many appear to believe, MSP is not a panacea. It provides measures that influence the spatial and temporal components of the oceans and how we use them. But, other measures and tools will be required that influence the performance of human activities and ecological processes of the marine environment, for example, quota or limits on infrastructure.

The MSP process recognises that due to overlapping objectives, not all uses are compatible with one another and that competing uses of ocean space can have adverse impacts on users. As such, it is argued that MSP can reduce conflicts, particularly those which are place- or area-based (Peel and Lloyd, 2004, Yates and Schoeman, 2013, Potts, 2015, Pomeroy and Douvere, 2008). MSP attempts to address conflict by helping to identify opposing interests, which may be between the same resource users (between fishers using different fishing gears); different resource users (e.g. between fishers and divers); or between different management agencies. It then enables these users to participate in defining (negotiated) desired goals and objectives regarding how the marine and coastal environment should be used, both now and into the future.

Over the last two decades, MSP has matured from a concept to an operational approach to contribute towards achieving sustainable development in the oceans. MSP initially arose out of the marine conservation agenda, and the first example of what was to become MSP could be seen in the zoning plans for the Great Barrier Reef in Australia in the 1980s. The discussion around the concept of MSP became global at the start of the twenty-first century, with an explosion of publications in the academic literature, as well as many policy documents and handbooks. Since this time, the uptake of the MSP process has been substantial with initiatives currently being undertaken around the world.

Using MSP to prevent conflict

Approximately 70 countries now have MSP processes in place. As of early 2019, 25 countries have marine spatial plans completed and 18 are government approved, 15 are currently being implemented, and seven are in the revision phase. This provides us with many opportunities to investigate how MSP processes have been used to resolve conflict. Here I present three examples from Europe, Asia, and North America.

Scotland, United Kingdom

The Scottish sea area comprises 61 per cent of the entire United Kingdom's sea area, covering 468,994 km^2 (out to 200 nautical miles), with a coastline that is 18,672 km long, and that incorporates 118 inhabited islands and over 800 uninhabited islands (Baxter *et al.*, 2011). These seas are home to a third of the global population of grey seals; the world's most northerly population of bottlenose dolphins; 23 other species of the world's 82 whales and dolphins; and 43 per cent of all seabirds breeding in the EU (ibid.). They provide summer feeding grounds for the basking shark (the world's second largest fish), and the endangered leatherback turtle (ibid.). Scottish seas also host rocky reefs covered in soft corals, sea fans, and sponges; shallow beds of delicate maerl, sea grass, flameshell, and horse mussel; as well as rare cold-water coral reefs (ibid.). However, these are all threatened by human activity.

In Scotland, MSP was initially trialled between 2006–10 as part of the Scottish Sustainable Marine Environment Initiative (SSMEI) project. The aim of the SSMEI was to gain an understanding of the nature, value, and management needs of Scotland's marine environment, and to develop and test new approaches to improve the sustainable management of Scotland's oceans and coasts. Four pilot areas were involved: The Firth of Clyde, the Shetland Islands, the Sound of Mull, and the Berwickshire coast. All pilots except for the Berwickshire coast, investigated the potential for, and in some cases developed, marine spatial plans.

In 2010, The Marine (Scotland) Act was adopted, allowing the Scottish Government unprecedented powers to plan its seas, except for some reserved functions, such as defence, which remain legislated by the UK Government. This act requires that Scottish Ministers prepare a National Marine Plan (NMP). As such, the Scottish Government has, between 2010 and 2015, undertaken an MSP process. The process comprised several steps. First, an assessment of the marine environment was carried out to create Scotland's Marine Atlas, and a Statement of Public Participation was published. Pre-consultation, as well as assessments to support the plan were then carried out, such as a sustainability appraisal, and the draft Plan was developed and published. Finally, several statutory processes were completed such as analysis of the consultation; an independent investigation of the plan, and Scottish and UK Ministerial clearance of the final Plan (Smith and Jentoft, 2017). The NMP was finalised in 2015 and follows the common vision of the UK Administrations as set out in the UK Marine Policy Statement and sets the broad aim of having 'clean, healthy, safe, productive, and biologically diverse oceans and seas' (HM Government, 2011: 3).

The NMP is to be implemented at a regional level through 11 Scottish Marine Regions. Within each region, delegated groups are responsible for drafting local marine plans; commonly referred to as Marine Planning Partnerships. To date, 3 of 12 plans have been completed: in Clyde and Shetland Islands regions (unsurprising, given their focus in the SSMEI), and the Orkney Islands region.

The Scottish MSP process appears to have contributed towards preventing marine and coastal conflict in two key ways. First, through the development of the Scottish Marine Atlas. This document presents a summary of the condition of the marine environment as it relates to the broad UK aim, as well as an assessment of human activity in Scotland's marine environment. In terms of human activity, data was collected on: aquaculture, commercial fishing, recreational fishing, leisure and recreational use, historic environmental and cultural heritage, coastal protection and flood defence, renewable energy, carbon capture and storage, oil and gas (and pipelines), water abstraction, maritime transport, waste disposal, military defence, communications, education, and research. An understanding of how and where the marine environment was being used, as well as an understanding of the pressures of these uses on the receiving environment, means that this Atlas provides the evidence base to help develop marine planning and management in a way which considers potential user-user and user-environment conflict.

In addition, during the preparation of the NMP, there were several occasions for stakeholder engagement: an initial 'joint workshop' to establish the plan's scope, content and objectives; pre-consultation on the draft; and consultation on the draft. However, as noted by Smith and Jentoft (2017) the stakeholder engagement for the NMP began with a single 'joint workshop' at which the majority of attendees had economic interests in decisions on the use of marine space. Those present included sectors such as: Fishing, Aquaculture, Ports and Harbours, Energy, Conservation, Defence, Leisure and Recreation, Local Authorities, Historical Assets, Tourism, and Shipping. Smith and Jentoft point out that these stakeholders have been staking their claims to marine resources for decades (for centuries in the case of sea fisheries), and power relations between them have been established, tried, tested, and re-established countless times. Furthermore, the interests of these stakeholders are not necessarily in harmony with those of coastal communities, who did not partake in the process. Similarly, there was little involvement by environmental non-governmental organisations. During the public consultation process, concerns were raised that the NMP focused too heavily on economic development, and too little on conservation of the marine environment or the impacts of climate change.

There is still the potential for conflict here. Perhaps this will be addressed further at the regional planning scale.

Savu Sea, Indonesia

Indonesia is the largest archipelagic nation in the world with a coastline of 95,181 km and 19,868 km^2 of the world's coral reefs (Perdanahardja and Lionata, 2017). At the southern end of the Coral Triangle, the Lesser Sunda Ecoregion stretches from Bali to Timor-Leste, covering an area of more than 45 million hectares and linking the Indian and Pacific Oceans. The region supports diverse and highly productive reef and pelagic habitats and is a major migratory corridor for cetaceans. Although the islands of the region are sparsely populated (*c.*13 million), human pressures include destructive fishing and overfishing, the illegal harvesting of cetaceans and turtles, and the disposal of mine tailings in the ocean (ibid.). The Savu Sea is a small sea within this region, named for the island of Savu on its southern boundary. Due to unique features such as deep trenches, strong currents, and areas of upwelling, the Savu Sea is an important habitat and migratory corridor for cetaceans and turtles, a key nursery area for fish, and a refuge for coral reefs. In 2008, Indonesia declared the creation of the 3.5 million-hectare Savu Sea Marine National Park.

The Nature Conservancy and Institut Teknologi Bandung have led the effort of engaging the Indonesian government in MSP. To begin the process of establishing the Savu Sea Marine National Park, an initial assessment of the area was undertaken (using the Nature Conservancy's ten-step Conservation Action Planning Process), identifying critical threats to the Savu Sea such as heat stress, marine debris, fishing practices, and ship strikes. This assessment suggested that zoning, as well as specific management actions around improved fishing practices would be an effective strategy for addressing conservation stress. The next step was a series of scientific surveys which looked at biodiversity features and threats, identified enabling conditions, and priority areas and seasons for patrolling. All data was then entered into a Marxan analysis to develop a zoning design. The final plan consisted of four types of zoning: a core zone, restricted to research and education; a sustainable fisheries zone; a marine tourism zone; and an 'others zone', further sub-divided into areas for traditional use, cetaceans, and cultural tourism.

In this case, the stakeholder engagement process was the key means of conflict prevention. The Nature Conservancy was involved with the design and implementation of the MPA network from its inception, and assisted a working team to assess, design, and establish the Savu Sea MPA. As part of the process, TNC was involved in the formation of P4KKP (Team for Assessment, Establishment and Management of Savu Sea National Marine Park) – a multi-stakeholder team for the preparation and establishment of the marine park. This team brought together over 1,000 people to discuss the development of the plan and was Indonesia's largest and most varied involvement of stakeholders in a natural resource decision-making process (Perdanahardja and Lionata, 2017). They also held public consultations in 94 villages. Public considerations were taken into account. In one community, there was a unique request to continue to harvest coral once every year for 'Keruga', or lime for the local betel nut delicacy. This request was later integrated into the management plan, allowed for subsistence but not commercial purposes.

This group has continued to establish forums and foster communications between agencies and stakeholders (including community leaders and civil society organisations) involved in park management. The collaborative mechanism in Savu Sea MPA is intended to be a continuous problem-solving process, involving extensive debates, negotiations, and joint learnings within problem-solving networks.

The MSP process is ongoing for the entire Lesser Sunda ecoregion.

Rhode Island, United States

In 2006, Rhode Island's governor committed to wind energy as an opportunity to generate economic growth while simultaneously tackling rising greenhouse gas emissions and energy costs. Rhode Island's biggest wind asset lay in its coastal and offshore waters. This drive for offshore wind energy extraction led to an MSP effort now known as the Rhode Island Ocean Special Area Management Plan (Ocean SAMP, McCann 2010); a process and plan largely aimed at minimising conflicts between this new use and traditional uses of the space.

Rhode Island's offshore waters are ecologically unique; shallow, near shore continental shelf waters which contain and host an interesting biodiversity of fish, marine mammals, birds, and sea turtles. Furthermore, the natural beauty of these waters, along with their rich historic and cultural heritage, provides aesthetic, artistic, educational, and spiritual value. This has meant that the waters off Rhode Island's coasts are an important and highly valuable environmental, economic and cultural hub for the people of this region. Commercial and recreational fishing, transportation, and tourism are just some of the ocean and coastal activities which have generated significant economic benefits for the region.

The Ocean SAMP planning process began in 2008 and while it began as an effort to proactively plan for wind farm development, it has since extended well beyond that focus on broader management of the diverse activities happening within state and federal waters. The process is a collaboration between the Rhode Island Coastal Resources Management Council (CRMC) and the University of Rhode Island alongside various other stakeholder groups. The Ocean SAMP has four goals: foster a properly functioning ecosystem that is both ecologically sound and economically beneficial; promote and enhance existing uses; encourage marine-based economic development that considers the aspirations of local communities and is consistent with and complementary to the state's overall economic development, social and environmental needs and goals; and build a framework for coordinated decision-making between state and federal management agencies. It is also underpinned by five principles: develop the Ocean SAMP document in a transparent manner; involve all stakeholders; honour existing activities; base all decisions on best available science; establish monitoring and evaluation that supports adaptive management.

The process began with scientific data collection (and new research where necessary) to provide knowledge on the Ocean SAMP area. At the same time, a stakeholder process was undertaken to provide local/traditional knowledge and to identify concerns. This information was formed into a draft document detailing the social-environmental properties of the area and presenting policies to promote and enhance existing uses and honour existing activities, which were then reviewed by stakeholders and technical advisory committees. The document was then put out to formal public comment and revised based on information received. The document was approved as part of the Rhode Island Coastal Resources Management Program and is now in a process of ongoing implementation and revision.

Similar to the Scottish example, in terms of addressing conflict, two key components are clear. First, to identify how the area had been, and was being used, data was collected regarding the various uses of the Rhode Island ocean and coastal waters. Data on the historic and cultural resources within Rhode Island's ocean and coastal waters were collected, recognising that the significance, sensitivity, and non-renewable nature of such resources and the special services they provide make them a challenging and important aspect of the Ocean SAMP process. Such data included the history and culture of indigenous Narragansett People, colonial settlement and the post-colonial cultural landscape, the military landscape, historic commercial and recreational fishing, historical shipwrecks and harbour features. There was also a strong focus on collecting data regarding commercial and recreational fisheries, recreation and tourism, marine transportation and energy production, as well as considering potential future uses of the environment such as mining, aquaculture, fisheries enhancement, and ecotourism.

In addition to the collection of data regarding the various and potentially conflicting uses of Rhode Island waters, those who ran the process recognised that development and implementation of the plan would also depend upon collaboration and engagement by all stakeholders. In this case, stakeholders were defined as government, citizens, civic and environmental organisations, resource users, and the private sector. To address this, the Ocean SAMP established an engagement framework which consisted of four components. The first component was the Ocean SAMP Stakeholder Group which played an integral role in determining the scope and contents of the plan, ground-truthing new research findings and refining the policies. A Technical Advisory Committee was established for each Ocean SAMP chapter (each area of focus in the plan) and was made up of scientists, government agency representatives, and resource users with expertise in the chapter topic. The Science Advisory Task Force included scientists representing different areas of scientific expertise. Finally, the Federal and State Agency Coordination contributed towards determining and responding to the scope of the Ocean SAMP document.

The first biennial assessment of the Ocean SAMP identified some ways in which conflict had been addressed through the process (Mulvaney, 2013). First, it was noted that a 13 square-mile renewable energy zone had been identified in state waters that directed development to a location which had the least conflict between existing uses and the natural environment. It was also pointed out that, for the first time, the existing activities within the Ocean SAMP area have been documented and that this would enable the proactive identification of potential conflicts. However, it was also clear that conflict hadn't always been addressed as the report raised that there was ongoing tension and a lack of collaboration between state agencies involved in the decision-making process.

MSP and the causes of conflict

These examples make clear that in some marine spatial processes, at least, there has been a focus on obtaining data regarding the different human activities operating in any given ocean and coastal environment. While this may help to identify where conflicting uses of the space may occur, how does the MSP process prevent/reduce/resolve conflict? I now examine how this tool addresses the exacerbating factors of conflict identified in Chapter 6 (for an overview, see Table 9.1).

Don't take/alter what is ours

Ocean zoning, which tends to be a key tool used as part of an MSP process, is believed to reduce conflict (Norse, 2002, Olson, 2010, Agardy, 2010). Ocean zoning is a means by which to divide a marine area into

Table 9.1 Overview of if, and how, marine spatial planning addresses the exacerbating causes of conflict

Exacerbating factor	*Sub-factor*	*Does MSP address (yes/no)?*	*Reason*
Don't take/alter what is ours	Exclusion	Yes and no	If zoning is used as part of MSP (which it often is), that creates exclusion. This can mitigate operational conflicts but unlikely to mitigate conflict over who has rights/precedence to operate in a given area.
	Subtraction	Yes and no	Can address some issues of subtraction such as aesthetics, but not others such as pollution.
Don't treat us unfairly	Distributional justice	No	Allocates use across time and space, unlikely that everyone will get what they want, therefore some may feel distribution of resources is 'unfair'.
	Procedural justice	Yes and no	Processes for procedural justice can be incorporated into the process but in reality are not.
Don't threaten our wellbeing	Individual	No	MSP process does not require a consideration of the downstream implications of allocation decisions.
	Community	No	MSP process does not require a consideration of the downstream implications of allocation decisions.
Don't govern/manage us poorly	Engagement and communication failure	Yes	Stakeholder engagement is a key component of the MSP process.
	Loss of trust	No	MSP cannot address trust directly. However, given that stakeholder engagement is a key component, if done correctly it is likely that trust will be gained.

districts and within those districts regulate use to achieve specified purposes. It is argued that this reduces user conflicts by separating incompatible activities and allocating or distributing uses based on a determination of an area's suitability for those uses. This supports the idea that exclusion minimises conflict.

While this may be true regarding operating conflicts (i.e. it may prevent offshore wind farms being installed in busy shipping lanes), it does not address the fact that the process of exclusion itself can lead to conflict (as identified in Chapter 6). Future conflict may be prevented, but only if agreement on zones can be reached by all stakeholders during the MSP process. Indeed, a study on offshore renewable energy development in Scotland suggested that to address conflict in a MSP process, the issues of who/what takes precedence, ownership, and rights in the marine space, as well as liability, must be addressed in such a process (Alexander *et al.*, 2012). Furthermore, conflict can only be prevented if all parties abide by agreements regarding zones of use and prohibited use.

MSP can prevent, reduce, or resolve conflicts caused by subtraction in some cases, but not in others. For example, by taking into consideration aesthetic value during the planning process, ocean and coastal development (which may be considered unsightly) can be sited in areas of lower natural beauty, thus reducing the potential for subtraction. However, due to the fluid and dynamic nature of the marine environment, it is less likely that MSP can easily address e.g. pollution, which can transfer between zones. As an example, eutrophication in the Baltic Sea has been a topic of contention between states bordering the space, as nutrient emissions drift from estuaries and coastal waters (particularly from Russia and Poland) to the Baltic proper and into the territorial waters of other countries (Tynkkynen *et al.*, 2014). Lines drawn on maps cannot stem the flow of contaminants. For this reason, MSP processes must include both zoning and other tools for managing the marine and coastal space.

Don't treat us unfairly

To date, insufficient attention has been paid to the wide range of potential distributive impacts that are caused and condoned by an uncritical rationalist approach to MSP (Flannery *et al.*, 2016). As part of the MSP process, human activities are distributed across space and time; distribution is therefore a key component of the process. However, can MSP ensure for the 'fair' allocation of adequate access to grounds (e.g. for mining, renewable energy extraction or conservation practices) and to stocks (e.g. for fishers) and what does this even mean? Furthermore, rarely do we ask who should be given access to marine resources and who should benefit from them. Yet surely this should be a key consideration in any MSP process.

As pointed out in Chapter 6, it is more likely to be the 'fairness' of the processes through which the distribution has occurred and the ability to avoid or ameliorate risk, or to access environmental resources, which are important. In this sense, the MSP process should, in theory, be able to afford the opportunity to address issues of justice and equity. However, this does not often appear to be the case, for example in Scotland where only certain stakeholders were included in the decision-making process.

As noted by Flannery *et al.* (2016: 126), while questioning who benefits from MSP:

> At its best, MSP could offer us a progressive, evidence-based process for guiding a fair exploitation of marine resources, thus helping to safeguard the wealth of our oceans for future generations. At worst it could be seen as a legitimating process for the continued market-led exploitation of marine resources, which powerful interests can manipulate for their own ends, obfuscating the real politics of development and the environment.

Don't threaten our wellbeing

MSP has the potential to ensure that individuals and local communities do not feel that their wellbeing is being threatened. However, the implications of making certain allocation decisions (on individuals and communities) are not required to be explicitly considered within the process and have not been reported in any marine spatial plan that I have seen. Even in the step-by-step process for MSP as adopted by UNESCO (Ehler and Douvere, 2009), only the section on defining and analysing future conditions comes anywhere near close by considering what the marine management area may look like under any specific MSP scenario.

It is almost certain that MSP will impact on certain sectors of society, particularly if they are not included in the decision-making process. It could cause a loss of subsistence, an impact on livelihoods, or a loss of identity. As an example, mobile fishing gears may not be compatible with fixed or floating infrastructure (such as wind farms or fish farms). As such, they are likely to be designated to operate in separate areas as part of an MSP process. However, in such cases, these locations may previously have been productive grounds, close to where fishers live, meaning that they have to travel further to alternative sites leading to increased competition with other fishers and increased spend on fuel, meaning fishers then have to spend longer fishing to break-even, leading to a downward spiral of losing time and money (Alexander *et al.*, 2013). This is a common fear for fishers regarding MSP.

Culture and identity tend to be a consideration within MSP in that they are enveloped within the frame of cultural ecosystem services. From this perspective, MSP should be able to consider community wellbeing. However, this is strongly linked to the section above regarding the inclusion, exclusion, or marginalisation of certain groups within the decision-making process. There is no requirement for consideration of the downstream implications of decisions for community wellbeing.

Don't govern/manage us poorly

It has been argued that people are central to the MSP decision-making process, are the agents for change and as such, stakeholder participation and involvement is integral to the success of MSP (Pomeroy and Douvere, 2008). Furthermore, stakeholder participation in MSP may help to address conflict by (i) identifying the compatibility or conflict of multiple use objectives and (ii) identifying, predicting and resolving conflict (ibid.). Indeed, in all three examples described in the previous section, we can clearly see a focus on stakeholder engagement. As such, this factor appears to be well considered within MSP processes.

Around the globe we can see a range of types of potential stakeholder participation in MSP from communication, where there is no actual participation, to negotiation, where decision-making power is shared among the various stakeholders (Pomeroy and Douvere, 2008). Participation is highly contested and while there are examples of how public engagement has led to progressive outcomes, there are also many cases of 'participatory' processes being executed in a superficial, top-down way with few benefits accruing to less powerful stakeholders (Flannery *et al.*, 2016), as is clear from the examples provided in Chapter 6 and in the previous section. This should, perhaps, act as a word of warning to those designing an MSP process.

In a study of what makes a good participatory process, Webler *et al.* (2001) proposed five perspectives: (i) that a good process acquires and maintains popular legitimacy; (ii) a good process is one that facilitates an ideological discussion; (iii) a good process should be fair; (iv) a good process should recognise power struggles; and (v) a good process involves good leadership and compromise. It may be that the inclusion of such principles into any MSP framework will address this exacerbating factor, thus reducing the potential for conflict. Furthermore, as argued by Pomeroy and Douvere (2008), stakeholder participation is critical, but it may not be adequate for the MSP process. Rather, stakeholder empowerment, through environmental education, capacity development and social communication, is essential and should be an integral component. Empower people with knowledge and skills and they can actively participate in MSP while simultaneously increasing their awareness and understanding of the marine environment and management.

Summary

MSP is a process of analysing and allocating the distribution of human activities in marine areas, across time and space, to achieve agreed-upon ecological, economic, and social objectives. The MSP process recognises that not all uses are compatible with one another and that competing uses of ocean space can have adverse impacts on users. As such, it has been argued that MSP can help to identify potential conflicts, and to address them by enabling users to participate in negotiations regarding how the marine and coastal environment should be used, both now and into the future. In this chapter, we have seen three examples of MSP processes, from the United Kingdom, Indonesia, and the United States. We have seen the ways in which they have attempted to address existing and potential conflicts.

In this chapter we also examined if, and how, MSP can address those exacerbating factors which can potentially tip conflict over the edge. I explained that if zoning is used as part of the process (which it often is), that leads to exclusion and that while this may mitigate operational conflicts it is unlikely to mitigate issues around rights/precedence. MSP can also address some issues of subtraction such as aesthetics, but not others such as pollution. Given the focus of MSP on allocating resource use across time and space, it is unlikely that everyone will get what they want. This is likely to be exacerbated by the fact that although the process could incorporate mechanisms for procedural justice, they often don't. Threats to well-being is one aspect which is unlikely to be considered by the MSP process. Finally, stakeholder engagement is a key component of the MSP process, and if the elements of a good participatory process are borne in mind, there is the potential that this exacerbating factor may be addressed.

MSP is a relatively new governance process within the ocean and coastal environment, with many countries currently working through the process. It is a tool which shows great potential for starting to address some of the factors which may lead to conflict. However, we should be wary of jumping in head-first, hoping that this will solve all our problems. MSP is not a panacea. It may be that we are still too early in the process to really identify some of the issues around power and influence that we are starting to identify in relation to other ocean and coastal governance mechanisms, such as co-management (discussed in more detail in Chapter 10).

References

Agardy, T. 2010. Marine spatial planning in the US. *Ocean Zoning: Making Marine Management More Effective*. Earthscan.

Alexander, K. A., Janssen, R., Arciniegas, G., O'Higgins, T. G., Eikelboom, T., and Wilding, T. A. 2012. Interactive marine spatial planning: Siting tidal energy arrays around the Mull of Kintyre. *PloS one*, 7, e30031.

Alexander, K. A., Potts, T., and Wilding, T. A. 2013. Marine renewable energy and Scottish west coast fishers: Exploring impacts, opportunities and potential mitigation. *Ocean & Coastal Management*, 75, 1–10.

Baxter, J. M., Boyd, I. L., Cox, M., Donald, A., Malcolm, S., Miles, H., Miller, B., and Moffat, C. 2011. *Scotland's marine atlas: Information for the national marine plan*. Marine Scotland.

Ehler, C. and Douvere, F. 2009. *Marine spatial planning: A step-by-step approach toward ecosystem-based management*. Intergovernmental Oceanographic Commission and Man and the Biosphere Programme.

Flannery, W., Ellis, G., Ellis, G., Flannery, W., Nursey-Bray, M., Van Tatenhove, J. P., Kelly, C., Coffen-Smout, S., Fairgrieve, R., and Knol, M. 2016. Exploring the winners and losers of marine environmental governance/Marine spatial planning: Cui bono?/'More than fishy business': epistemology, integration and conflict in marine spatial planning/Marine spatial planning: power and scaping/Surely not all planning is evil?/Marine spatial planning: a Canadian perspective/Maritime spatial planning–'ad utilitatem omnium'/Marine spatial planning: 'it is better to be on the train than being hit by it'/Reflections from the perspective of recreational anglers and boats for hire/Maritime spatial planning and marine renewable energy. *Planning Theory & Practice*, 17, 121–51.

HM Government. 2011. *UK Marine Policy Statement*. Her Majesty's Stationery Office.

IOC-UNESCO. n.d. *MSP at IOC-UNESCO* [Online]. Available: http://msp.ioc-unesco.org/about/msp-at-unesco/ [Accessed 20 February 2019].

Mulvaney, K. K. 2013. First biennial assessment of the Rhode Island Ocean Special Area Management Plan Process. Prepared for the Rhode Island Coastal Resources Management Council and the University of Rhode Island Coastal Resources Center. Funded by the Rhode Island Office of Energy Resources. Available: https://seagrant.gso.uri.edu/oceansamp/pdf/documents/doc_osamp_evaluation.pdf [Accessed 11 March 2019].

Norse, E. A. 2002. A zoning approach to managing marine ecosystems. *Improving Regional Ocean Governance in the US*, 53–7.

Olson, J. 2010. Seeding nature, ceding culture: Redefining the boundaries of the marine commons through spatial management and GIS. *Geoforum*, 41, 293–303.

Peel, D. and Lloyd, M. G. 2004. The social reconstruction of the marine environment: towards marine spatial planning? *Town Planning Review*, 75, 359–78.

Perdanahardja, G. and Lionata, H. 2017. *Nine years in Lesser Sunda. Indonesia: The nature conservancy*. Indonesia Coasts and Oceans Program.

Pomeroy, R. and Douvere, F. 2008. The engagement of stakeholders in the marine spatial planning process. *Marine Policy*, 32, 816–22.

Potts, T. 2015. Marine spatial planning and various uses and interests relating to the marine environment. *Transboundary Marine Spatial Planning and International Law*. Routledge.

Smith, G. and Jentoft, S. 2017. Marine spatial planning in Scotland. Levelling the playing field? *Marine Policy*, 84, 33–41.

Tynkkynen, N., Schönach, P., Pihlajamäki, M., and Nechiporuk, D. 2014. The governance of the mitigation of the Baltic Sea eutrophication: Exploring the challenges of the formal governing system. *Ambio*, 43, 105–14.

Webler, T., Tuler, S., and Krueger, R. 2001. What is a good public participation process? Five perspectives from the public. *Environmental Management*, 27, 435–50.

Yates, K. L. and Schoeman, D. S. 2013. Spatial access priority mapping (SAPM) with fishers: A quantitative GIS method for participatory planning. *PloS one*, 8, e68424.

Zaucha, J. and Gee, K. 2019. *Maritime spatial planning: past, present, future.* Springer.

Chapter 10

Co-managing conflict

In Chapters 8 and 9, I examined how two governance mechanisms – laws and marine spatial planning – addressed (or did not address) those factors which have the potential to aggravate conflict. A third governance mechanism, which will be investigated in this chapter, is the process of co-operative management, or co-management.

In the past, some traditional societies managed natural resources through complex interplays of reciprocities and solidarities which were fully embedded in local cultures and meaning systems. For other societies, social values such as religious authority, caste predestination and cultural norms formed the basis of natural resource decisions. The historical emergence of colonial powers, however, and their assumption of authority over natural resources often led to the demise of such traditional natural resource management systems. The monetisation of resources weakened local systems of reciprocity and solidarity, a demise which was accelerated by the move towards a global economy. In addition, the rise in power of modern, science-based decision-making induced losses of local knowledge and skills. All these factors combined have led to the disempowerment of local communities in natural resource management (Borrini, 2000).

More recently, governance failures associated with a propensity for top-down management arising from such changes in natural resource management have spawned a variety of institutional arrangements to engage resource users. One such tool, emerging from the idea of involving citizens in making decisions about their own livelihoods, and reducing implementation costs and improving compliance (Viswanathan, 2003), is co-management (Ayers and Kittinger, 2014). In this chapter, I will explore the potential for co-management – as an approach to marine and coastal governance which may be used to address marine and coastal conflict.

In this chapter, I provide an overview of co-management. I present some case examples where co-management has helped to address conflict. Last but not least, I assess how co-management can address the four exacerbating factors: our rights (or lack of) to common resources, our needs for social justice, threats to our wellbeing, and poor governance.

Co-operative management of the oceans and coasts

The Organisation for Economic Co-operation and Development (OECD) defines co-management as 'a process of management in which government shares power with resource users, with each given specific rights and responsibilities relating to information and decision-making' (Organisation for Economic Co-operation and Development, 2008). The World Bank defines co-management as 'the sharing of responsibilities, rights and duties between the primary stakeholders, in particular, local communities and the nation state; a decentralized approach to decision-making that involves the local users in the decision-making process as equals with the nation-state' (World Bank, 1999: 11). Given the many definitions of co-management it is clearly a concept which does not appear to be well-defined and because it means different things to different people it is potentially subject to a spectrum of interpretations. Yet, as explained by Carlsson and Berkes (2005), the definitions and conceptualisations of co-management, although coming from a variety of sources, do have some common underpinnings. They explicitly associate the concept of co-management with natural resource management. They regard co-management as a partnership between public and private actors. They stress that co-management is not a fixed state but a process that takes place along a continuum. Moreover, co-management is not about government delegating all responsibility for core functions. The responsibilities retained by government tend to involve powers to make regulations; powers to initially grant rights to access resources; and powers of compliance, investigation, and prosecution.

In addition to the many definitions of co-management, several typologies of co-management have been proposed. Sen and Nielsen (1996) suggest five types of co-management form along a spectrum: (i) instructive management where the government instructs resource users as to what laws and policies they are required to follow and information is shared with the community towards the end of the planning process; (ii) consultative management where the government consults with the community either through public hearings or advisory boards but is not required to implement any of the suggestions or comments into the policy; (iii) co-operative management where there is a degree of co-management of the resource by the government and the user group and responsibility is shared by both parties; (iv) advisory management where users decide what decisions should be made and advise the government as such and the government than endorses the decision; and (v) informative management where the user group makes all of the decisions and informs the government once they have decided. Like Sen and Neilsen, McCay (1996) and Pomeroy and Berkes (1997) present co-management arrangements which also appear to follow Arnstein's (1969) ladder of public participation.

Since the 1990s co-management schemes, whereby the management of resources is shared by public and/or private sector stakeholders, have been a key strategy for reducing natural resource conflicts worldwide. Some believe that co-management can foster community empowerment, ensure inclusive decision-making and legitimacy and lead to benefit sharing and, ultimately, livelihood enhancement (Soliku and Schraml, 2018). Others argue that co-management is all about empowerment; bringing previously excluded, disenfranchised, and sometimes alienated stakeholders into the decision-making process, by reshuffling power and responsibility among those who manage the relevant natural resource (Jentoft, 2005). Nonetheless, there has been much debate regarding the effectiveness of such strategies. It has been suggested that co-management often glosses over the institutional complexities posed by the management of common pool resources, that it empowers the elite in the background while purporting to facilitate participation, and that in some cases, co-management has exacerbated the conflict that it was intended to solve (De Pourcq *et al.*, 2015). Co-management can be successful in reducing conflict at grass-roots level, if some critical enabling conditions are fulfilled not only on paper but also by praxis (ibid.). Co-management can also help mitigate or prevent conflicts from escalating (Soliku and Schraml, 2018).

Conflict can also be a substantial motivating factor for stakeholders to become involved in co-management approaches. As noted by Viswanathan (2003), co-management approaches can increase the authority held by specific communities which may then be used by that community to exclude outsiders from accessing 'their' natural resources.

Using co-management to prevent conflict

Co-management is an approach which has been used across a wide variety of natural resource management spaces. For example, in marine protected area management (e.g. Avalos and Torero, 2015), water management (e.g. Conallin *et al.*, 2017), and in the case of disaster risk reduction (e.g. Nguyen and Tran, 2016). However, the scholarly literature would suggest that, in the marine and coastal sphere at least, the approach can be found most frequently in the marine fisheries space. Indeed, one systematic review of small-scale fisheries around the globe found 91 cases of co-management spread across 37 countries, in Europe, the Americas, Africa, Asia, and Oceania (d'Armengol *et al.*, 2018). In providing examples of how co-management has been used to address conflict, the following three examples focus on conflicts relating to fishing.

Destructive vs. non-destructive fishing conflicts, Philippines

San Salvador is an island village (*barangay*) located in the province of Zambales in the Philippines. It is located 250 km north of Manila and requires a 30-minute boat ride to get to the island. The 380 hectares of San Salvador island are surrounded by white-sand beaches, except for the east coast which comprises barren forest and mud flats. Reefs, dominated by seagrass and algae can be found off the northern, western, and southern coasts. Mangrove reforestation efforts were undertaken in 1990 but most of the 8,000 seedlings were destroyed due to mudflow from Mt. Pinatubo (Katon *et al.*, 1997).

Fishing and farming are the dominant economic activities in San Salvador; the fishery is multi-gear, multi-species, and largely artisanal (Kuperan *et al.*, 1999). Fishers argue that prior to the Second World War because fish were abundant there was no need for destructive fishing practices, in fact there was no competition for resources. During the Second World War, Japanese soldiers occupied the island and sometimes used dynamite to catch fish (ibid.). In the 1970s, San Salvador saw an influx of Visayan immigrants searching for better fishing grounds, and this time saw the extreme shift to destructive fishing practices such as blast fishing, the use of sodium cyanide and spear fishing with compressors. These practices eventually decimated San Salvador's fishing grounds (ibid.). They also gave rise to conflicts over fishing gear and over productive fishing spots between the Visayans on the one hand, who used sodium cyanide and dynamite, and the rest of the residents, who normally deployed non-destructive fishing gear such as hook-and-line, nets, and spear guns (Katon *et al.*, 1997, Katon *et al.*, 1999).

Fostered by a de facto open-access fishery and lack of law enforcement, illegal fishing practices prevailed. The situation could neither be controlled by the distant highly centralised national government of the Philippines nor by the fragmented San Salvador fishers (Katon *et al.*, 1999). However, in March 1987, a Peace Corps volunteer working in the Bureau of Fisheries and Aquatic Resources arrived in San Salvador. The volunteer initiated discussions with several stakeholder groups, including municipal government and non-governmental organisations (NGOs), and through this gained support for rehabilitating the fishery resources of the island – leading to consideration of a marine sanctuary in 1988 (Kuperan *et al.*, 1999).

Conceived as a community-based coastal resource management (CBCRM) project, the Haribon Foundation (one of the first NGOs to recognise the important role of the community in resource management projects in the Philippines) was selected as the lead implementing organisation (Katon *et al.*, 1999). The project was funded by the Netherlands Embassy, the Jaime V. Ongpin Foundation, and the World Wildlife Fund's Debt for Nature Swap Program (Kuperan *et al.*, 1999). The programme was then developed and implemented by the Haribon Foundation until 1993, when it was turned over to the people's organisation that it helped to establish – the Samahang Pangkaunlaran ng San Salvador (ibid.).

The aim of the project was to enhance capabilities (of the NGOs and the municipal government), develop and implement a marine resource management plan, and establish a coral reef fish sanctuary and marine reserve. The process also encouraged the formation and strengthening of local groups responsible for marine resource management, using intensive information campaigns and educational site visits including a ten-day trip to Apo Island, a marine sanctuary in the Central Philippines (Christie *et al.*, 1994). In 1989, the core group members led a campaign to create the 127-hectare marine sanctuary and reserve of San Salvador, drafting a local ordinance that banned fishing within the sanctuary and allowed only non-destructive fishing methods in the marine reserve (Katon *et al.*, 1999), meaning that rights of access (entry rights) and withdrawal (harvesting) were created where previously there were none. The Masinloc Municipal Council then passed an ordinance (Municipal Resolution No. 56, Series of 1989) to provide legitimacy to the marine sanctuary and reserve at the local level. The municipal government was also involved in the co-management process in various other ways: mediating conflicts; providing enforcement-related resources such as a boat, radios and fuel, and the implementation of a patrol team; and the creation of a political environment that allowed for community-based initiatives (Katon *et al.*, 1999).

The key outputs of co-management were the development of community organisations, including an Environment Management Committee and an Alternative Income Committee. The responsibility of the former is to develop plans, introduce them to the community and to work with the community to ensure their engagement in and support of the plans. The Alternative Income Committee is responsible for working with small groups to develop project plans for income generating projects. The development of these groups, which play key roles in the process, has provided a strong grass-roots component to the co-management process (Christie *et al.*, 1994).

In an analysis of the outcomes of the San Salvador co-management process, fishers perceived increases in knowledge of fisheries, information exchange, satisfaction with fishery arrangements, benefits from the marine reserve, and faster resolution of community conflicts (Katon *et al.*, 1997, Katon *et al.*, 1999).

Artisanal vs. semi-industrial fishing conflicts, Mozambique

In Mozambique, artisanal fishing has been important in the socio-economic life of the country for the social stability it provides among the coastal communities. According to a study undertaken in 2007 by the

World Wildlife Fund, artisanal fishing employed 80,000 fishermen distributed among 787 fishery centres, where 2,781 artisanal fishing vessels were licensed, fishing in near-shore areas mostly using beach seine methods, gillnets, or handlines. Fishers operate using *lanchas* (also known as a 'dhow' which has a characteristic triangular lateen sail), small boats locally known as *almadias*, and open-mouthed canoes (Santos, 2007). Semi-industrial fisheries also operate along the Mozambique coast, employing approximately 1,550 fishers, fishing for shrimp in the deeper waters using refrigeration vessels (ibid.). Prawn fishing has come to be considered one of the most important national revenue generators, accounting for approximately 40 per cent of the total exports of Mozambique (Lopes and Gervásio, 2003).

A high concentration of beach seines can be found along the coast of the Angoche and Moma districts, an area which is at the northern tip of the Sofala Bank (the continental shelf), which is also a key area for the semi-industrial trawls (Viswanathan, 2003). This has led to conflicts between the two fisheries. On one hand, the semi-industrial trawls operate very closely to the shore leading to the destruction of the artisanal beach seine gear. On the other hand, the semi-industrial fishers claim that artisanal fishers are catching many larvae and juvenile fish (including juvenile shrimp) due to their smaller mesh sizes, and that this is impacting the shrimp fishery (ibid.).

In terms of management of such conflicts, prior to independence, control of fishing areas and other natural resource remained the responsibility of the local traditional authority, considered the 'real' owners of such property. The traditional chief held the authority to approve (or not) negotiations for the allocation of fisheries in their territory (Lopes and Gervásio, 2003). This changed after independence and it is not clear how this management system was replaced between this time and 1990 when the Fisheries Act was first adopted as Act no. 3/90. Given that Mozambique was in a time of civil war between 1975 and 1990, it is unlikely that any fisheries legislation was adopted. Cultural rituals prevailed, and these can be considered a traditional form of resource management at the community level. These include thanksgiving ceremonies known as Olomba and Sataka in which participants ask for blessings of their gods and surveillance over the perceived destroyers of their resources (Lopes and Gervásio, 2003).

The Fisheries Act established the framework for fisheries (particularly the industrial fisheries), and other laws were also established in 1990 to fill some of the gaps, particularly around licensing and enforcement (Afonso, 2004). The Fisheries Master Plan was approved in 1996 and laid out the priorities and strategies for development. It was in this plan that the concept of co-management was introduced with an emphasis on the involvement of fishermen in setting and enforcing management regimes (Lopes and Gervásio, 2003, Viswanathan, 2003).

Co-management has been implemented at the national level. For example, a fisheries management committee (the *Commisao de Administracao Pesquiera*) was established which included representatives from the artisanal and semi-industrial sectors (Viswanathan, 2003). It has also been implemented at the district and local levels, with the agreement in Moma-Angoche being one of the best-known examples. Co-management in this region was facilitated by the Nampula Artisanal Fisheries Project which ran from 1994 to 2001. The project promoted the formation of local Resource Management Committees and Co-management Committees with the aim of developing fisheries management mechanisms to be implemented in close collaboration between the fishermen and the Maritime Administration (IFAD, 2000). Scholars investigating the role of co-management in Mozambique have focused particularly on co-management in Kwirikwidge, a rural fishing village near the town of Angoche, and an area in which artisanal fishing had (by the turn of the millennium) reached commercial level.

In 1997, the National Institute for the Development of Small-scale Fisheries (IDPPE) began a process of working with community fishers aimed at achieving greater participation of these groups in identifying problems as well as decision-making. This was done by creating resource management committees at the local and district levels. The roles of these committees are: to inform fishers about the need to use fishing materials and equipment that enable sustainable use of marine resources; to discuss and make decisions around problems/conflicts resulting from fishing; and to refer issues to the relevant authorities where necessary (Lopes and Gervásio, 2003). In addition, an inter-district level committee was also created, where representatives of all community committees, including Kwirikwidge, were able to meet with the responsible government agency (ibid.).

Research into the effectiveness of co-management processes identified that there had been a considerable reduction in the rates of conflict between artisanal fishers resulting from the theft of gear or the entanglement of nets. There had also been a reduction in the number of disputes between native and immigrant fishers. Regarding conflicts between the artisanal fishers and the semi-industrial fleet, the local committee, in coordination with the inter-district committee worked together to identify ways of preventing such conflicts, resulting in a 3-mile trawler exclusion zone (Lopes and Gervásio, 2003).

Co-Management of Coastal Fisheries in Hawai'i

The Hawaiian Islands are an archipelago of eight major islands, several atolls, numerous smaller islets, and seamounts in the North Pacific Ocean, extending some 2,400 km from the island of Hawai'i in the south to northernmost Kure Atoll. Hawai'i is the most isolated population centre in the world and is also one of only two US states with tropical coral reefs. The islands were first settled by Polynesians and marine resources were managed by *konohiki* (land administrators appointed by local chiefs). Konohiki worked to gather input from local fishers and *maka'āinana* (land tenants) to develop place-based rules that governed resources sustainably (Kirch, 2019). Since annexation by the United States, marine management has changed substantially, however, since the 1980s, the state has seen a renewal of traditional ecological knowledge and governing institutions (Ayers, 2016), and a move towards co-management processes.

Communities in Hawai'i rely substantially on fisheries for economic, social, and cultural services, including important livelihood and food provisioning functions. This is particularly true for near shore coral reef-associated fisheries which support a range of commercial, recreational, and subsistence activities, employing multiple gear types, and harvesting a wide variety of reef and estuarine finfishes, invertebrates, and schooling coastal pelagic species (McCoy *et al.*, 2018). Yet, the fisheries harvests have declined substantially over the last 100 years, with several stocks depleted below benchmark sustainability levels (ibid.).

As such, the islands have seen increased competition over resources, caused by this depletion. Outsiders are entering areas previously only accessed by nearby communities, and technological advances have enabled an influx of new fishers to enter previously inaccessible or remote fishing areas. Disputes over resource access became protracted conflict over rights to marine resources. In some instances, communities took enforcement into their own hands using intimidation or on occasion illicit actions to protect their marine resources (Ayers and Kittinger, 2014). Conflict caused by resource depletion has been a key driver in the move to co-management approaches in Hawai'i, combined with minimal State enforcement of rules governing fisheries.

Community concern regarding the regulatory regime has led to consideration of alternative management options, one such option being the development of community-based subsistence fishing areas (CBSFAs) in addition to management partnerships between State agencies and local communities (Kittinger *et al.*, 2012). CBSFAs allow fishing communities to partner with the Hawai'i Division of Aquatic Resources to create rules based upon the customary and traditional Native Hawaiian use of their marine resources. In response to a recommendation made by the Governor's Subsistence Task Force and sustained lobbying efforts made by the Hui Mālama o Mo'omomi community organisation and others, Act 271 was passed by the Hawai'i Legislature in 1994. This gave the State of Hawai'i's Department of Land and Natural Resources the power to designate CBSFAs and establish a two-year pilot project for Mo'omomi and Kawa'aloa Bays on the island of Moloka'i (Ayers, 2016). Community-developed rules for the pilot project included spatial measures defining permitted and prohibited fishing activities, gear restrictions, a permitting system managed by the DAR, and training of community volunteers to monitor activities and educate novice fishers (ibid.). Unfortunately, the pilot project was allowed to expire at the end of two years (Kittinger *et al.*, 2012). While several communities expressed interest in crafting CBSFAs, the Pilot Project represented the only CBSFA in Hawai'i until 2015 when Hawai'i's governor approved rules for a CBSFA in Hā'ena, Kaua'i (Ayers *et al.*, 2017).

Despite this, communities across Hawai'i began to self-organise, to educate, to monitor resources and to build consensus around solutions. Examples of this include: the Kaua'i community of Hā'ena working with the State Park managers and other entities to increase local involvement in Hā'ena coastal area management in response to increased tourism and recreation; and the Miloli'i community responding to increased fishing pressure from outsiders by lobbying and testifying in support of a bill to permanently designate their adjacent marine areas as a CBSFA, although ultimately the community was unable to reach a consensus. (Ayers and Kittinger, 2014). The difficulties in obtaining community consensus was believed to be a key challenge to co-management practices, particularly when, as was the case for the Miloli'i community, proposals were developed by a limited number of individuals (Ayers *et al.*, 2017).

Despite such hurdles, however, there is still great interest in co-management across the state. A policy briefing note by Kittinger *et al.* (2012) suggests that areas under active community management have revealed ecological benefits in respect to fish biomass and social benefits such as the increased engagement of disparate stakeholder groups, incorporation of traditional ecological knowledge and customary practices into decision-making, and enhanced learning and capacity building.

Co-management and the causes of conflict

Co-management, as a practice which requires various entities to jointly manage a resource, needs to be able to address conflict to function. The examples above show that this can happen in a variety of ways: through the creation of new organisations, agencies, and communities working together to draft legislation, or by empowering individuals to enable knowledge transfer and social learning. But can co-management get to the heart of those exacerbating factors identified in Chapter 6? This is what I investigate in this section (for summary overview see Table 10.1).

Don't take/alter what is ours

Co-management is not in itself about property rights but development of a co-management regime requires (and facilitates) determination of access to

Table 10.1 Overview of if, and how, co-management addresses the exacerbating causes of conflict

Exacerbating factor	*Sub-factor*	*Does co-management address (yes/no)?*	*Reason*
Don't take/alter what is ours	Exclusion	Yes	Although co-management is not in itself about property rights, development of a co-management regime requires (and facilitates) determination of access to resources.
	Subtraction	Yes	As part of the agreed-upon co-management process, users are allocated subtraction rights.
Don't treat us unfairly	Distributional justice	Yes and no	Co-management occurs along a spectrum. Even at most devolved end of spectrum, distributional justice may not occur. Theoretically it provides potential mechanism for distributional justice.
	Procedural justice	Yes and no	Co-management occurs along a spectrum. Even at most devolved end of spectrum, procedural justice may not occur. Theoretically it provides potential mechanism for procedural justice.
Don't threaten our wellbeing	Individual	Yes	Co-management not required to consider downstream effects on individual wellbeing. However, has been found to be beneficial to wellbeing.
	Community	Yes	A key concept behind co-management is to cede responsibility to communities to manage their own resources and to make decisions affecting their well-being.
Don't govern/ manage us poorly	Engagement and communication failure	Yes	As communities of interest are involved directly in process, there should be no failure of engagement. Knowledge transfer and social learning are key components of co-management.
	Loss of trust	Yes	Co-management requires the working together of various stakeholders under certain power sharing arrangements. This, in turn, requires trust-building.

resources. Co-management builds upon a larger set of property options for managing natural resources, including various forms of community-based jurisdiction over natural resources, or at least rights to use and manage them. Co-management does not require any specific ownership system and as such there are examples of co-management working within many different property rights regimes (Jentoft *et al.*, 1998). In fact, it has been argued that co-management steers a middle course between pure state property and pure communal property regimes (Pomeroy and Berkes, 1997). This has meant that often, as we have seen in the examples, the community works with government to determine property rights, in whatever form may be most appropriate, which are then enshrined in law to provide legitimacy. As the resource users are most often involved in co-management, it should logically follow that they have agreed upon who can access what, and as such shouldn't feel aggrieved when this happens, unless of course others are not obeying the rules. The same should hold true also for the issue of subtraction.

Don't treat us unfairly

In theory, co-management should provide a potential mechanism for social justice. The argument goes that co-management leads to empowerment, which in turn leads to social justice (Jentoft, 2005). As suggested earlier, co-management occurs along a spectrum. Yet even at the most devolved end of spectrum, procedural and distributional justice may not occur. In discussions around the meaning of co-management, we see words like 'collaborate', 'share', and 'cooperation', qualities that Davis and Ruddle (2012) note are desirable for social justice. These authors also explain, however, that the presumption that the neoliberal state is willing to share authority to empower marine harvesters and enhance social justice is problematic. Furthermore, as found by Gustavsson *et al.* (2014), co-management can create new participation and distributional injustices caused by elite capture (whereby public resources are biased for the benefit of a few individuals) when decentralised management reinforces existing power structures within the community and creates new opportunities within the decentralised administration. None of the examples provided earlier discussed social justice outcomes.

Don't threaten our wellbeing

Co-management is not required to consider downstream effects on individuals. However, a meta-analysis which focused on small-scale fisheries (Evans *et al.*, 2011) demonstrated that co-management resulted in positive impacts on fishers' income and other sources of material well-being. A systematic review of small-scale fisheries by d'Armengol *et al.*

(2018) resulted in similar findings regarding wellbeing. Such findings are not clear, however, in the examples provided earlier.

When it comes to community wellbeing, a key concept underpinning co-management is to cede responsibility to communities to manage their own resources and to make decisions affecting their wellbeing, indicating that wellbeing is central to use of the approach. The cases examined earlier appear to evidence this, particularly from the perspective of community culture where top-down government approaches were often combined with bottom-up traditional management approaches based on customary practice, ensuring that rights and rituals remained a part of the co-management practice. It was not clear how this part of the process was undertaken. Of course, there will be challenges in reconciling differing worldviews such as those of indigenous and Western management approaches (Levine and Richmond, 2014). There is also the potential that specific elements of traditional ecological knowledge are 'cherry-picked' for integration with western science, and that they are increasingly divorced from the broader social and cultural context in addition to those that hold the knowledge (Stevenson, 2004).

Don't govern/manage us poorly

Co-management is not so much about the rules as about the communicative and collaborative process through which rules are formed. Co-management is all about who participates, how debates are structured, how knowledge is employed, how conflicts of interest are addressed, and how agreements are reached (Jentoft *et al.*, 1998). As such, there should be no failure of engagement. As noted previously, this factor relates closely to that of procedural justice and there are several potential pitfalls in this space. As Jentoft *et al.* (1998) note, there is a basic problem of who will get a 'seat at the table' of co-management. Stakeholder groups, such as fishers, are not a homogenous entity; within these communities exist hugely divergent worldviews. The difficulty is, of course, that the more participants are involved, the harder it is to find the acceptable middle ground. Knowledge transfer and social learning are key components of co-management, perhaps this may be a way in which to ensure engagement, without all having a 'seat at the table'.

Regarding issues around trust, co-management requires the working together of various stakeholders under certain power sharing arrangements. This requires trust-building. Trust builds up through repeated interaction and institutionalisation of the links made (Adger *et al.*, 2005). It is important to be aware, however, that powerful agents, with exclusive knowledge and lots of influence, can undermine trust in collaborative working arrangements (ibid.).

Summary

Marine and coastal co-management, or the joint management of the marine commons is most often understood as some arrangement of power sharing between the State and a community of resource users. Although the approach has been used across a wide variety of natural resource management situations it is used most frequently regarding management of marine fisheries, particularly those that are small-scale or artisanal. Such an approach is becoming a key strategy for reducing natural resource conflicts worldwide, although there is some debate regarding just how effective co-management can be.

Three case study examples revealed a range of approaches to co-management, even though all were dealing with fisheries conflicts. In Asia, Africa, and America, the development of community organisations was key, as was the legitimacy provided by the backing of State entities. In two of the three cases it was also noted that the process had helped to reduce and resolve conflict.

In this chapter, I argued that co-management is an approach which has the potential to address many of the factors which exacerbate marine and coastal conflict. Although co-management is not in itself about property rights, development of a co-management regime requires (and facilitates) determination of exclusion and subtraction of common resources. It provides a mechanism for procedural and distributional justice, although there are several potential pitfalls. Co-management has been found to be beneficial to both individual and community wellbeing, although again, there are many considerations around how to integrate traditional and western cultures. Finally, engagement and trust-building are central to co-management, but we must remember that power and influence in participatory processes can be problematic.

Of the three mechanisms I have analysed in relation to the factors which exacerbate marine and coastal conflict, co-management appears to provide the most comprehensive coverage of the factors. I want to make clear, however, that this does not mean that I am suggesting that this is the 'best' mechanism. A toolbox always comprises a variety of hammers, spanners, and wrenches, and marine and coastal governance is no different. We should always apply the right tool for the job.

References

Adger, W. N., Brown, K., and Tompkins, E. L. 2005. The political economy of cross-scale networks in resource co-management. *Ecology and Society*, 10.

Afonso, P. S. 2004. *Country review: Mozambique* [Online]. FAO. Available: www.fao.org/3/a0477e/a0477e10.htm [Accessed 23 February 2019].

Arnstein, S. R. 1969. A ladder of citizen participation. *Journal of the American Institute of Planners*, 35, 216–24.

Avalos, F. M. and Torero, M. G. 2015. Challenges to sustainable development along Peruvian coastal zones. *Coastal Zones: Solutions for the 21st Century*, 200–11. Elsevier.

Ayers, A. 2016. *From planning to practice: Toward co-management of Hawai'i coral reef fisheries*. University of Hawaii at Manoa.

Ayers, A. L. and Kittinger, J. N. 2014. Emergence of co-management governance for Hawai'i coral reef fisheries. *Global Environmental Change*, 28, 251–62.

Ayers, A. L., Kittinger, J. N., Imperial, M. T., and Vaughan, M. B. 2017. Making the transition to co-management governance arrangements in Hawai'i: A framework for understanding transaction and transformation costs. *International Journal of the Commons*, 11, 388–421.

Borrini, G. 2000. *Co-management of natural resources: Organising, negotiating and learning-by-doing*. Kasparek Verlag.

Carlsson, L. and Berkes, F. 2005. Co-management: concepts and methodological implications. *Journal of Environmental Management*, 75, 65–76.

Christie, P., White, A. T., and Buhat, D. 1994. Community-based coral reef management on san Salvador island, the Philippines. *Society & Natural Resources*, 7, 103–17.

Conallin, J. C., Dickens, C., Hearne, D., and Allan, C. 2017. Stakeholder engagement in environmental water management. *Water for the Environment*. Elsevier.

D'Armengol, L., Castillo, M. P., Ruiz-Mallén, I., and Corbera, E. 2018. A systematic review of co-managed small-scale fisheries: Social diversity and adaptive management improve outcomes. *Global Environmental Change*, 52, 212–25.

Davis, A. and Ruddle, K. 2012. Massaging the misery: recent approaches to fisheries governance and the betrayal of small-scale fisheries. *Human Organization*, 71, 244–54.

De Pourcq, K., Thomas, E., Arts, B., Vranckx, A., Léon-Sicard, T., and Van Damme, P. 2015. Conflict in protected areas: Who says co-management does not work? *PloS one*, 10, e0144943.

Evans, L., Cherrett, N., and Pemsl, D. 2011. Assessing the impact of fisheries co-management interventions in developing countries: A meta-analysis. *Journal of Environmental Management*, 92, 1938–49.

Gustavsson, M., Lindström, L., Jiddawi, N. S., and De La Torre-Castro, M. 2014. Procedural and distributive justice in a community based managed marine protected area in Zanzibar, Tanzania. *Marine Policy*, 46, 91–100.

IFAD. 2000. *Nampula Artisinal Fisheries Project* [Online]. Available: www.ifad.org/en/web/ioe/evaluation/asset/39827939 [Accessed 23 February 2019].

Jentoft, S. 2005. Fisheries co-management as empowerment. *Marine Policy*, 29, 1–7.

Jentoft, S., McCay, B. J., and Wilson, D. C. 1998. Social theory and fisheries co-management. *Marine Policy*, 22, 423–36.

Katon, B., Pomeroy, R., Salamanca, A., and Garces, L. 1997. The marine conservation project for San Salvador: A case study of fisheries co-management in the Philippines. *Fisheries Co-management Research Project Working Paper*, 23, 10–95.

Katon, B. M., Pomeroy, R. S., Garces, L. R., and Salamanca, A. M. 1999. Fisheries management of San Salvador Island, Philippines: A shared responsibility. *Society & Natural Resources*, 12, 777–95.

Kirch, P. V. 2019. *How chiefs became kings: Divine kingship and the rise of archaic states in ancient Hawai'i*. University of California Press.

Kittinger, J., Ayers, A., and Prahler, E. 2012. Policy Briefing. Co-management of coastal fisheries in Hawai'i: Overview and prospects for implementation. *Co-Management of Coastal Fisheries in Hawai'i: Overview and Prospects for Implementation*, 23 October.

Kuperan, K., Abdullah, N., Pomeroy, R., Genio, E., and Salamanca, A. 1999. Measuring transaction costs of fisheries co-management in San Salvador Island, Philippines. *Naga, the ICLARM Quarterly*, 22, 45–8.

Levine, A. and Richmond, L. 2014. Examining enabling conditions for community-based fisheries co-management: Comparing efforts in Hawai'i and American Samoa. *Ecology and Society*, 19.

Lopes, S. and Gervásio, H. 2003. Co-management of artisanal fisheries in Mozambique: A case study of Kwirikwidge Fishing Centre, Angoche District, Nampula Province.

McCay, B. J. 1996. Participation of fishers in fisheries management. Fisheries resource utilization and policy: *Proceedings of the World Fisheries Congress*, 60–75. Science Publishers.

McCoy, K. S., Williams, I. D., Friedlander, A. M., Ma, H., Teneva, L., and Kittinger, J. N. 2018. Estimating nearshore coral reef-associated fisheries production from the main Hawaiian Islands. *PloS one*, 13, e0195840.

Nguyen, H. and Tran, P. 2016. Urban disaster risk reduction in Vietnam: Gaps, challenges, and approaches. *Urban Disasters and Resilience in Asia*. Elsevier.

Organisation for Economic Co-operation and Development 2008. *OECD glossary of statistical terms*. OECD.

Pomeroy, R. S. and Berkes, F. 1997. Two to tango: The role of government in fisheries co-management. *Marine Policy*, 21, 465–80.

Santos, R. 2007. *Fisheries in Angoche, Moma and Pebane: A preliminary description*. WWF.

Sen, S. and Nielsen, J. R. 1996. Fisheries co-management: A comparative analysis. *Marine Policy*, 20, 405–18.

Soliku, O. and Schraml, U. 2018. From conflict to collaboration: The contribution of co-management in mitigating conflicts in Mole National Park, Ghana. *Oryx*, 1–11.

Stevenson, M. G. 2004. *Decolonizing co-management in Northern Canada* [Online]. Cultural Survival. Available: www.culturalsurvival.org/publications/cultural-survival-quarterly/decolonizing-co-management-northern-canada [Accessed 24 February 2019].

Viswanathan, K. K. 2003. *Fisheries co-management policy brief: Findings from a worldwide study*. WorldFish.

World Bank 1999. Report from the international workshop on community-based natural resource management (CBNRM).

Chapter 11

Stemming the tide

Growing disputes over the appropriation and exploitation of the oceans and coasts have meant that since the 1980s, a variety of marine- and coastal-related legal and policy instruments have emerged to address the effects of various human activities taking place within the marine environment. These have led to a shift in the paradigm from situations where no regulation or limited governance have existed to a more heavily regulated and managed environment. Yet conflict continues, despite the increase in marine governance institutions and mechanisms.

At the very start of this book, I explained that my aim was to 'discern conflict and contestation in the marine and coastal environment' and 'what it means for sustainable development, conflict theory and practice'. The first was answered in Chapter 7. Now I will synthesise the findings detailed in this book to answer the second. In this chapter, I provide a brief recap of all previous chapters. I then address what the findings of this book mean for the policies of blue growth and conservation specifically, within the framework of the sustainable development goals. Finally, I provide some implications for theory and, conceivably more importantly, for practice in the prevention of marine and coastal conflict.

Chapter summary

Chapter 1 introduced us to some examples of marine and coastal conflict, and to the problem being addressed in this book: that of why conflict occurs in this environment and what it means for the potential for future sustainable development of the marine space. I also explained the process of analysis used in this study; providing some background to the results provided later in the book.

The study of marine and coastal conflict proper started in Chapter 2 where I explored the differences between marine and terrestrial environments. I claimed that the physical environments were different, that the ecosystem services provided were different, that the 'common goods' nature of these ecosystem services meant that they had to be managed differently, and that we held a very different relationship with the marine environment from that with the terrestrial. I argued that these differences would be relevant to how we understand conflict in one environment as opposed to the other.

In Chapter 3, I delved into the concept of conflict, the scholars that have attempted to understand what causes social and natural resource conflict, and their theories. I showed that existing theories of social and environmental conflict do not meaningfully engage with some of the most important aspects of the marine environment, particularly the physical environment, the 'commons' nature of ecosystem services and the contested nature of the space. I explained that it is for this reason that a separate treatment of marine and coastal conflict is required.

Prior to starting the process of offering a theory of marine and coastal conflict, I presented several case studies in Chapter 4. These were taken from across the globe, as much as possible, and covered a broad range of marine and coastal conflicts relating to a variety of ocean and coastal resource conflicts including water, fish, oil and gas, wind and wave energy, mangroves, saltmarshes, and amenity. These case studies provided the basis for a grounded theory analysis of the causes of marine and coastal conflict.

The results of this analysis were presented in Chapters 5 and 6. In the former, I identified that five drivers were forcing competition over marine resources. These included: an increasing global population; the need for water, food and energy security; the imperative for ongoing economic growth; a failure in governance; and a changing marine environment. I argued that competition can be good or bad, that it is always present, but it can be transformed into conflict. In the latter chapter, I identified four factors that may turn competition into conflict: rights to ocean and coastal resources, perceived fairness, threats to wellbeing, and poor governance. I argued that competition, when it becomes personalised, can lead to conflict and that while competition is likely to continue, we can prevent/reduce/resolve marine and coastal conflict by considering these 'personalised' factors – ones which can be changed.

Chapter 7 took a brief sojourn back into the world of conflict theory, laying down explicitly what I believe to be the first theory of marine and coastal conflict. I argued that increased competition for ocean and coastal goods and services, which is driven by a changing social-ecological environment, is transformed into conflict by political and social factors that shape the rules, rights, and effects of human resource use. I explained that there are clear overlaps with existing theories of conflict but also some differences. I made clear that this is at the early stages of theory development, needing much further testing and cross-validation while providing us with a 'strawman' starting point to consider the management of conflict within current marine and coastal governance mechanisms – the focus of the following three chapters.

In Chapter 8 I identified that laws can have an important role to play in conflict prevention and resolution. Law seeks to address the factors proposed to exacerbate marine and coastal conflict in various ways, by determining the rights of users to access and exploit marine resources, by ensuring health and safety or protecting heritage and culture, and by including stakeholder engagement as part of the process of enacting legislation and part of the legislated process for decision-making. I argued that despite addressing some factors, we cannot rely on law along to prevent, reduce, or resolve marine and coastal conflict.

In Chapter 9, I introduced the concept of marine spatial planning (MSP), noting that this tool can help to identify potential conflicts, and address them through user-user negotiations. MSP addresses the exacerbating factors using zoning to address exclusion and subtraction, it could incorporate processes for procedural justice (and good stakeholder engagement) but often doesn't, and wellbeing is unlikely to be addressed by MSP. I argued that the tool shows great potential, but we may be too early in the process to fully understand how it can, and does, prevent, reduce, or resolve marine and coastal conflict.

Finally, in Chapter 10, I investigated the potential for co-management to address conflict, noting that it has become a key strategy around the globe for dealing with conflict. I argued that co-management is an approach with the potential to address many of the exacerbating factors. It requires determination of ocean and coastal resource rights, it provides a mechanism for procedural and distributive justice, it has been found to be beneficial for wellbeing, and engagement and trust-building are central to the process. I further argued that although co-management appeared to more comprehensively address the exacerbating facts, we should bear in mind that this doesn't mean we should always choose to use this tool to address conflict. We should remember that there are many tools in the toolbox.

Overall, the three governance mechanisms do start to address marine and coastal conflict, particularly the first exacerbating factor regarding our rights to use and alter the marine and coastal environment. They also have the potential to address many of the other factors. To do this, those undertaking marine and coastal management may need to shift their focus.

Implications for sustainable development of the oceans and coasts

The concept of a 'blue economy' came out of the 2012 Rio+20 Conference, emphasising conservation and sustainable management of the oceans and coasts, based on the premise that healthy marine ecosystems are more productive and a must for sustainable ocean-based economies. Growth in ocean-based economic sectors, so called 'blue growth', has come from improved access to, and more efficient utilisation of, ocean and coastal natural resources, and has become a key part of the development agenda of many coastal States. For example, the European Union has a Blue Growth Strategy (European Commission. Directorate-General for Maritime Affairs and Fisheries, 2012). Furthermore, to support the move towards the blue economy, the Food and Agriculture Organisation of the United Nations (FAO) have launched the 'Blue Growth Initiative' (Eikeset *et al.*, 2018), through which it will assist countries in developing and implementing blue economy and growth agendas.

As we move towards blue growth, we are simultaneously seeing an increase in the number of marine areas that are being set aside for conservation purposes. The Convention on Biological Diversity (CBD), the most important international legal instrument addressing protected areas, has set a target of achieving 10 per cent of coastal and marine areas set aside as 'effectively and equitably managed, ecologically representative and well-connected systems of protected areas and other effective area-based conservation measures' (Convention on Biological Diversity, 2012). In 2017, the CBD suggested that the world was on track to achieving this target, with the total area covered by Marine Protected Areas globally having increased nearly 20-fold since 1993 and having more than doubled since 2010 (Convention on Biological Diversity, 2017).

In this time of blue growth and increasing marine protection, we have also seen the establishment of the Sustainable Development Goals (SDGs), introduced in the introductory chapter. The SDGs provide a shared blueprint for peace and prosperity for people and the planet, now and into the future. They recognise that addressing some of the greatest challenges we face must go hand-in-hand with strategies that improve health and education, reduce inequality, and spur economic growth – all while working to preserve our oceans (and other environments). The SDGs provide an operating framework which ties together blue growth and marine conservation; enabling human development while at the same time sustaining the ability of natural systems to provide the natural resources and ecosystem services upon which we depend.

As we have seen, this increased access to, use of, and protection of resources, has led to conflict between sectors (e.g. tourism vs. offshore hydrocarbon extraction), at different levels of organisation (e.g. between individuals, groups, and nations), at multiple spatial scales (e.g. in local waterways, regional seas, or global oceans), and across time (e.g. between current and future uses). Continued blue growth in addition to increasing marine conservation, as we attempt to achieve the SDGs, is likely to lead to more conflict and many socially undesirable outcomes.

So, how can the findings of this study help us to address the move towards addressing the SDGs? First, it has been suggested that there is intrinsic conflict between sustainability and development in terms of resource allocation because they attempt to achieve two very different goals (Yamagata, 2015). While there may some truth in this, in this study I have shown that despite a superficial glance suggesting that marine and coastal conflicts represent this dichotomy, this is not what underpins the conflict. This would suggest that, if we can address the actual underlying factors leading to the perception of this conflict, these supposedly divergent goals should not subvert our ability to achieve the SDGs.

The 2018 version of the yearly Sustainable Development Goals Report found that conflict and inequality are major factors hindering progress towards the SDGs (United Nations, 2018). This report was referring to violent conflict and its effects on displacement and food security particularly. However, as we have seen, even small-scale, non-violent conflict can affect food and energy security, health and wellbeing, and impair the social contract between citizens and state. SDG 16 explicitly recognises the need to resolve conflict and mitigate its consequences but does not provide specific action points regarding how to do this. Perhaps the findings presented in this book can provide an opener to start to address conflict, at least in the marine and coastal environment.

Moreover, many of those factors which lead to conflict are directly related to some of the SDGs, particularly those relating to poverty (SDG 1); health and wellbeing (SDG 3); decent work and economic growth (SDG 8); inequality (SDG 10); as well as the goal relating to peace, justice and strong institutions (SDG 16) discussed above. By addressing the underlying causes of marine and coastal conflict, it may be that we can contribute towards achieving these SDGs.

Implications for theory and practice

The findings of this study highlight implications for conflict theory. First, it is not useful to directly apply existing theories of social or natural resource conflict theory when considering the marine environment. The marine and coastal environment is substantially different from the terrestrial, and therefore requires a separate consideration of conflict as has been presented here. Second, there is a broad range of types of marine and coastal conflict, both user-user and user-environment, regarding a variety of marine resources, each with its own context. Yet as we have seen, it is possible to draw generalisations which may be useful for considering the operationalisation of governance tools and approaches to prevent marine and coastal conflict. This is a 'young' theory and as such would benefit greatly from further testing and cross-validation. Furthermore, it may be interesting, from a theoretical perspective, to also test this theory against other terrestrial 'common pool' resources such as forests.

There are also several implications for practice. There are five key factors driving competition for marine and coastal resources: an increasing global population; the need for water, food, and energy security; the imperative for ongoing economic growth; a failure in governance; and a changing marine environment. They are often global in scale, and it is unlikely that we can change them, so it is not useful to consider such drivers in the context of addressing marine and coastal conflict. There are also four key 'exacerbating' factors which may turn this competition into conflict: rights to ocean and coastal resources, perceived fairness, threats to wellbeing, and poor governance. These are personal and tend to result in some sort of loss (e.g. of livelihood, health, wellbeing, culture, identity, potential for self-actualisation) for individuals or groups. They can more easily be addressed and as such should be our focus when designing new marine governance mechanisms to prevent marine and coastal conflict. Existing governance tools/approaches already (most likely inadvertently) address various exacerbating factors, but (at least for the mechanisms addressed here) never all of them. We have seen that in many instances there is potential within the mechanism to address these factors. Therefore, to prevent conflict arising after creating new legislation or a marine spatial plan, or establishing a co-management process, we must consider the four exacerbating factors during process design.

It is likely that this will involve thinking about the tools and approaches that we use in new and novel ways. If we take the 10-step guide to MSP, developed by Ehler and Douvere (2009) for UNESCO, as an example, we can start to see where we could consider the exacerbating factors. In step 1: identifying need and establishing authority, we can address the poor governance factor by being clear and transparent about roles and responsibilities. Rather than waiting until step 4 to involve stakeholders, why not considering introducing them in step 3: organising the process through pre-planning? We could address the fairness and poor governance factors by working with communities of interest to establish the defining principles, goals and objectives, and risks of the process. In step 5: defining and analysing existing conditions, rather than focusing purely on biology, ecology, and human activities, consider collecting data on health and wellbeing, as well as culture and identity, as relate to the areas under consideration, thus addressing the wellbeing factor. In step 7: preparing and approving the spatial management plan, make very clear who has rights to do what and where across the marine space, addressing the rights to resources factor. In step 9: monitoring and evaluating performance, include means by which to monitor how the exacerbating factors are being addressed. If they are not being addressed, adapt the process as is required in step 10. This provides just one example of how we can consider the exacerbating factors throughout the development of a governance process. There will be many other ways in which to do this.

Conflict prevention is the object of a wide range of tools and approaches in marine and coastal governance; aiming to avoid the escalation of a dispute. Despite the systemic conditions that make conflict prevention difficult, renewed thinking and focus on conflict prevention is urgent as competition over ocean and coastal resources continue to rise. After all, it is far more effective and less expensive (for all involved) to prevent conflict than to respond to it. As Desiderius Erasmus (1466–1536), perhaps the greatest scholar of the northern Renaissance, is attributed to have said: 'Prevention is better than cure'.

References

Convention on Biological Diversity. 2012. *TARGET 11 – Technical Rationale extended (provided in document COP/10/INF/12/Rev.1)* [Online]. Available: www.cbd.int/sp/targets/rationale/target-11/ [Accessed 24 February 2019].

Convention on Biological Diversity. 2017. *Global marine protected area target of 10% to be achieved by 2020*. Secretariat of the Convention on Biological Diversity.

Ehler, C. and Douvere, F. 2009. *Marine spatial planning: A step-by-step approach toward ecosystem-based management*. Intergovernmental Oceanographic Commission and Man and the Biosphere Programme.

Eikeset, A. M., Mazzarella, A. B., Davíðsdóttir, B., Klinger, D. H., Levin, S. A., Rovenskaya, E., and Stenseth, N. C. 2018. What is blue growth? The semantics of 'Sustainable Development' of marine environments. *Marine Policy*, 87, 177–9.

European Commission. Directorate-General for Maritime Affairs and Fisheries. 2012. *Blue growth: Opportunities for marine and maritime sustainable growth*. Communication from the Commission to the European Parliament, the Council, the European Economic and Social Committee and the Committee of the Regions, Publications Office of the European Union.

United Nations 2018. *The Sustainable Development Goals Report 2018*. United Nations.

Yamagata, T. 2015. *Sustainable development goals: The conflict between sustainability and development* [Online]. United Nations Economic and Social Commission for Asia and the Pacific. Available: www.unescap.org/sites/default/files/2015-S2-2%20Presentation%20Summary_Yamagata.pdf [Accessed 24 February 2019].

Index

For Product Safety Concerns and Information please contact our EU representative GPSR@taylorandfrancis.com
Taylor & Francis Verlag GmbH, Kaufingerstraße 24, 80331 München, Germany

www.ingramcontent.com/pod-product-compliance
Ingram Content Group UK Ltd.
Pitfield, Milton Keynes, MK11 3LW, UK
UKHW022124190926
475702UK00013B/1628
9780367785376